Alaska's First Bush Pilots, 1923–30

And the Winter Search in Siberia for Eielson and Borland

Alaska's First Bush Pilots, 1923–30

And the Winter Search in Siberia for Eielson and Borland

by Jim Rearden

Pictorial Histories Publishing Company, Inc.
Missoula, Montana

Library of Congress Control Number 2009938329

ISBN 978-1-57510-147-7

FIRST PRINTING October 2009
SECOND PRINTING September 2016

Unless otherwise credited, all photos in this book are from the Noel Wien Collection, courtesy of Richard Wien

TYPOGRAPHY & BOOK DESIGN Arrow Graphics

PRINTED BY Friesens Inc.
Altona, Manitoba, Canada

Published by Pictorial Histories Publishing Company, Inc.
713 South Third Street West, Missoula, Montana 59801
PHONE (406) 549-8488, FAX (406) 728-9280
E-MAIL: phpc@montana.com
WEBSITE: pictorialhistoriespublishing.com

Contents

Two-page Map of Alaska viii–ix
List of Place Names ix
Foreword x
Introduction xiii
Acknowledgments xvi

Book One: The Beginnings

1. Ben Eielson Arrives at Fairbanks 3
2. Noel Wien Arrives at Fairbanks 19
3. The Early Years of Wien Airlines 30
4. Alaska's First Bush Pilot 48
5. Olaf Swenson, Trader 62
6a. Map of Siberia 76
6. Flight to Siberia 77

Book Two: Eielson's Exploration Flights

7. Eielson and Wilkins' 1926 Flights 87
8. Eielson and Wilkins' 1927 Flights 99
9. The Flight to Siberia 105
10. Flights in Antarctica 116

Book Three: The Siberian Challenge

11. Travails of the Nanuk; The Hamilton Disappears 123
12. The Search for the Hamilton 139
13. Recovering the Bodies 156
14. Payback Time 173

Book Four: The Rest of the Story

15. Joe Crosson 181
16. Harold Gillam 200
17. Ed Young 215
18. Frank Dorbandt 232
19. Noel Wien, the Later Years 244
20. Key Players in the Eielson/Borland Saga 256

References *263*
Appendix *267*
Index *269*

Also by Jim Rearden

Alaska's Wolf Man
The 1915–55 wilderness adventures of Frank Glaser

Sam O. White Alaskan
Tales of a Legendary Wildlife Agent and Bush Pilot

Castner's Cutthroats
Saga of the Alaska Scouts

The Wolves of Alaska
A fact-based saga

Forgotten Warriors of the Aleutian Campaign

Koga's Zero
The Fighter That Changed World War II

Slim Moore: Alaska Master Guide
A Sourdough's Hunting Adventures and Wisdom

Jim Rearden's Alaska
Fifty Years of Frontier Adventure

Travel Air NC9084
The History of a 75-Year-Old Working Airplane

Hunting Alaska's Far Places
Fifty Years with Rifle and Shotgun

All thirteen of these books can be found in most Alaska book stores. The above ten may be ordered directly from the publisher (1-888-763-8530).

Arctic Bush Pilot
From Navy Combat to Flying Alaska's Northern Wilderness

Tales of Alaska's Big Bears

Shadows on the Koyukuk
An Alaskan Native's Life Along the River

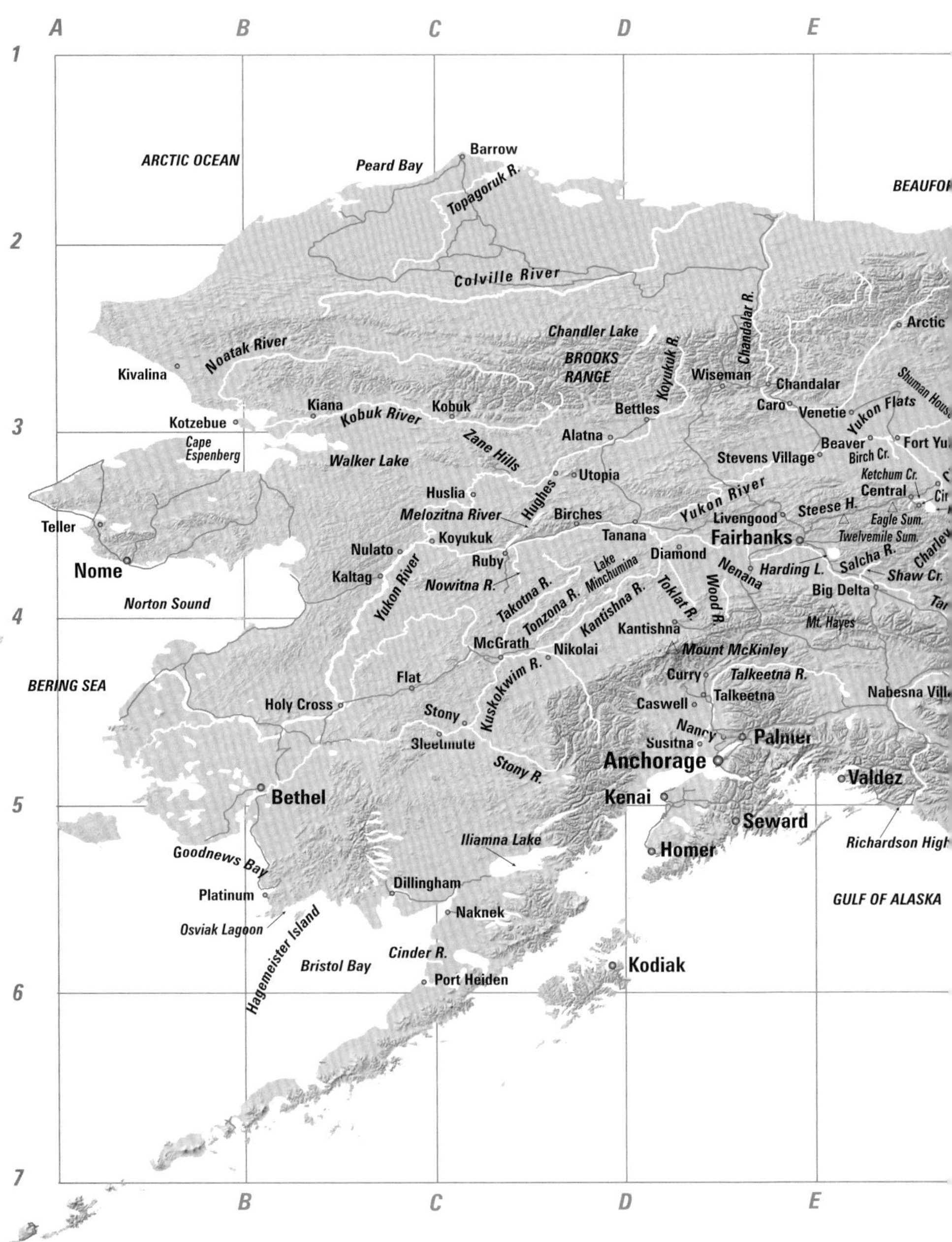

A
B
C
D
E
1
2
3
4
5
6
7
ARCTIC OCEAN
Peard Bay
Barrow
Topagoruk R.
BEAUFOR
Colville River
Chandler Lake
BROOKS RANGE
Noatak River
Kivalina
Koyukuk R.
Chandalar R.
Arctic
Wiseman
Chandalar
Shuman House
Kiana
Kobuk
Bettles
Caro
Venetie
Yukon Flats
Kotzebue
Kobuk River
Alatna
Beaver
Fort Yu
Cape Espenberg
Zane Hills
Walker Lake
Stevens Village
Birch Cr.
Utopia
Ketchum Cr.
Huslia
Hughes
Central
Yukon River
Melozitna River
Birches
Livengood
Steese H.
Eagle Sum.
Teller
Koyukuk
Tanana
Fairbanks
Twelvemile Sum.
Nulato
Diamond
Ruby
Lake Minchumina
Salcha R.
Nome
Harding L.
Kaltag
Nowitna R.
Nenana
Shaw Cr.
Norton Sound
Yukon River
Takotna R.
Tonzona R.
Kantishna R.
Toklat R.
Wood R.
Big Delta
Mt. Hayes
Kantishna
McGrath
Nikolai
Mount McKinley
BERING SEA
Flat
Kuskokwim R.
Curry
Talkeetna R.
Holy Cross
Talkeetna
Nabesna Vill
Caswell
Stony
Nancy
Palmer
Sleetmute
Susitna
Anchorage
Stony R.
Valdez
Bethel
Kenai
Seward
Iliamna Lake
Homer
Richardson High
Goodnews Bay
Dillingham
Platinum
GULF OF ALASKA
Naknek
Osviak Lagoon
Hagemeister Island
Cinder R.
Bristol Bay
Kodiak
Port Heiden

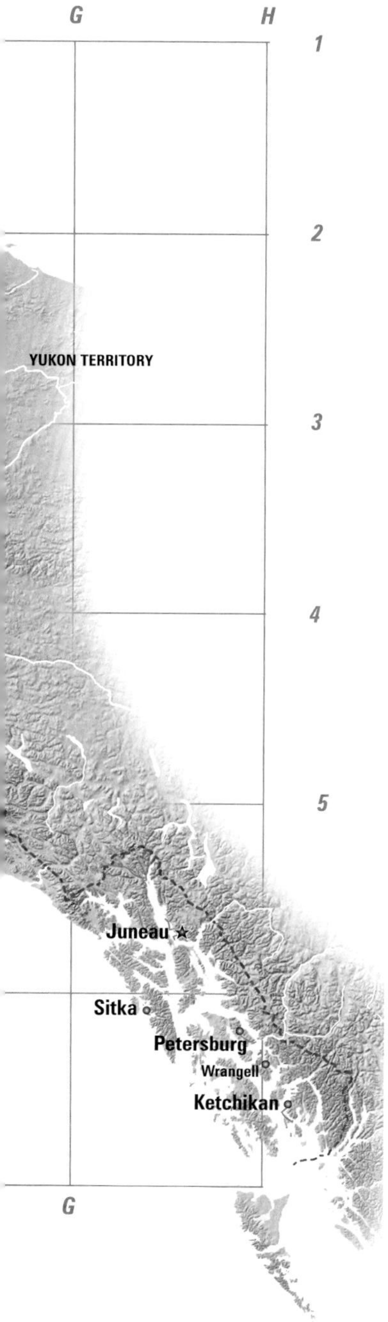

Place	Grid
Alatna	D3
Anchorage	D5
Arctic Village	E2
Barrow	C2
Beaver	E3
Bettles	D3
Big Delta	E4
Birch Creek	E3
Birches	D4
Black River	F3
Bristol Bay	B6
Brooks Range	D3
Cape Espenberg	B3
Caro	E3
Caswell	D4
Central	F3
Chandalar	E3
Chandler Lake	D2
Charley River	F4
Chatanika	E4
Chicken	F4
Cinder River	C6
Circle City	F3
Circle Hot Springs	F3
Coleen River	F3
Colville River	C2
Curry	D4
Diamond	D4
Dillingham	C5
Eagle	F4
Eagle Summit	E3
Fairbanks	E4
Flat	C4
Fort Yukon	E3
Goodnews Bay	B6
Hagemeister Island	B6
Harding Lake	E4
Holy Cross	B4
Howling Dog Rock	F3
Hughes	D3
Huslia	C3
Iliamna Lake	C5
Joseph Creek	F4
Joseph	F4
Juneau	H6
Kaltag	C4
Kantishna	D4
Kechumstuck	F4
Ketchikan	H7
Kiana	B3
Kivalina	B3
Kobuk	C3
Kobuk River	C3
Kotzebue	B3
Koyukuk River	D3
Koyukuk	C4
Kuskokwim River	C4
Ladue River	F4
Lake Minchumina	D4
Livengood	E3
McGrath	C4
Medicine Lake	F3
Melozitna River	C3
Mount McKinley	D4
Mt. Hayes	E4
Nabesna	F4
Naknek	C6
Nancy	D5
Nation River	F3
Nenana	E4
Nikolai	D4
Noatak River	B3
Nome	A4
Northway	F4
Nowitna River	C4
Nulato	C4
Osviak Lagoon	B6
Palmer	E5
Peard Bay	C2
Platinum	B6
Porcupine River	F3
Port Heiden	C6
Richardson Highway	B5
Ruby	C4
Salcha River	E4
Shaw Creek	E4
Shuman House	F3
Steese Highway	E3
Stevens Village	E3
Stony River	C5
Stony	C5
Susitna Station (Susitna)	D5
Takotna River	D4
Talkeetna	D4
Talkeetna River	E4
Tanana River	F4
Tanana	D3
Teller	A3
Tetlin	F4
Toklat River	D4
Tonzona River	D4
Topagoruk River	C2
Twelvemile Summit	E3
Utopia (Indian Mtn)	D3
Valdez	E5
Venetie	E3
Walker Lake	C3
Wiseman	D3
Wood River	D4
Wrangell	H6
Yukon Flats	E3
Zane Hills	C3

Foreword

IN THE EARLY DAYS of Alaska aviation, Fairbanks, being the geographic center of the Territory, became the natural service and supply point of the entire Interior, including the lower Yukon and Kuskokwim Rivers. Before the advent of the airplane, most commerce was handled by river travel in summer, and dog team in winter. When the first airplanes began replacing the river boats and dog teams, Fairbanks became the base of operations for the early pilots.

Starting around 1927, a major technological change developed in aviation, and it very much affected the dynamics in Alaska. It allowed my father, Noel Wien, in 1927 to initiate the first year-around regular air service between the two largest mainland Alaska communities, Fairbanks and Nome. This was made possible with the purchase of the Stinson Detroiter cabin biplane from arctic explorer Hubert Wilkins.

This Stinson was the first American-made cabin plane in Alaska. It moved the pilot out of the open cockpit that had long been in vogue, and which was impractical in Alaskan winters. Also, this airplane was powered by an air-cooled 220 hp Wright engine. The Wright was the first reliable engine that could function in the cold winters of Alaska's Interior. This was the same engine that powered the *Spirit of St. Louis* across the Atlantic. Prior to the Wright, aircraft in Alaska and elsewhere used unreliable liquid-cooled engines, which were especially unsuitable for Alaskan winters.

A superb, modern, air-cooled engine that followed the Wright was the famous 420 Pratt and Whitney Wasp. My father often talked about his 1929 flight to North Cape, Siberia, with his new Hamilton Metalplane, the first airplane to fly in Alaska with the

Wasp engine. It was an all-metal state-of-the-art, modern-for-its-time aircraft that he purchased in late 1928 for $26,000. That airplane, NC10002, played a major role in Alaska's aviation history. It was an airplane he loved and flew for more than 450 hours before he sold it to Alaskan Airways, the company managed by Ben Eielson.

As my brother Merrill and I grew up, Noel often talked about the fur trader Olaf Swenson, who chartered our Dad's Hamilton to make the first ever flight, North America to Asia, by flying to Siberia and the ice-locked trading ship *Elisif.* It held a cargo of valuable fur that needed to be flown to Fairbanks so it could be shipped via rail and sea to New York and London fur markets.

In recent years I discovered Swenson's book (*Northwest of the World*, Dodd Mead & Company, 1944). I was thrilled to read about Swenson's exploits, and his use of small ships to trade in Siberia, and the details of his being locked in ice at North Cape, Siberia. His story, too, is part of the Alaska-Siberian aviation epic.

I have also long been fascinated by the story of Ben Eielson, lost with his mechanic Earl Borland while attempting to fly the Hamilton my father had owned to Siberia and the iced-in *Nanuk*. It too held a cargo of valuable furs. The extensive midwinter search for the missing Hamilton and its two aviators was the most dramatic Alaska aviation epic of the period.

I have always been in awe of the flying of Joe Crosson and Harold Gillam in the winter of 1929-30 as they searched for the Hamilton. Having flown an open cockpit airplane, I cannot imagine flying in one as they did in temperatures of forty below zero, and with hardly any daylight. Gillam, who had just learned to fly, talked Crosson into letting him have an airplane to join the search. Other than the Hamilton and the Stinson Detroiter, there were no modern cabin aircraft in the Territory. At the time, the Stinson was damaged, leaving only open cockpit biplane aircraft available for the search. Fortunately, both search planes had Wright air-cooled engines.

Even today it would be difficult with a modern airplane with a heated cabin and radios to conduct a search under the arctic conditions these two pilots encountered.

My brother Merrill and I are often asked to make slide presentations about the early years of Alaska flying by our father and others. For these talks we have used many photos Noel took (many of which appear in this book). We both have vivid memories of his stories associated with these photos. In the process, we have frequently attempted to tell the story of the Siberian experience. It is a complex story, and it has been difficult to articulate the entire event during a brief evening presentation. We have both thought it important to have a book written that covered all aspects of the Eielson Siberian saga, including the background of early Alaska aviation and its pilots. It is a story that needed to be told in its entirety, which has finally become a reality with this volume.

I have known Jim Rearden for more than fifty years, and I have read many of his books. I helped him with his fine book *Sam O. White, Alaskan,* for Sam, an early-day flying game warden and bush pilot, was almost a second father to Merrill, me, and our sister Jean. It was then that Jim and I began to talk about the possibility of his writing about the Eielson/Borland saga. I gathered all of the information, books, and photos I had for him to study, and I was pleased when he agreed to tackle the project.

I have been impressed with Jim's style of writing, and his ability to capture the essence of people and events. With this book, recounting the struggles of Alaska's first bush pilots, he has done it again. I believe it will be one of the best of the many historical books on Alaska's early aviation.

—Richard Wien

Introduction

THIS BOOK IS A LOOK BACK at Alaska's earliest aviation, mostly at Fairbanks, where the first commercial flights were made by Ben Eielson with a World War I JN4 (Jenny) open cockpit biplane. In addition to its pilot, who flew from the rear cockpit, everything it could fly had to fit into that airplane's front cockpit.

Eielson was the first in Alaska's Interior to demonstrate that airplanes were more than a unique source of entertainment. Early commercial pilots, soon called "bush pilots," were looked upon as heroes; "bush pilot" is still an honorable title in Alaska.

For landing places, the first pilots used river bars, baseball fields, and race tracks (first used by horses, later by cars, finally by airplanes). Their airplane motors (the early name; "engines" today) were cranky and liquid-cooled. Power failures were common. Landings on rough ground often resulted in a broken propeller, damaged landing gear, a crushed radiator, and mangled tips of the lower wings of the biplanes they flew. Prepared pilots always carried an extra prop lashed to the side of the fuselage.

To fly, pilots must have visibility. There were no weather forecasts for the early birds. A pilot might depart under blue skies, and a hundred miles and an hour-long flight away be forced by fog or heavy snow to land. Airplanes had no radios. Once in the air, a pilot was out of touch, and on his own. In the early years it took nerve and self-confidence for a man to climb into the cockpit of a biplane and embark on a flight across the wild land that is still much of Alaska.

Airplanes were commonly forced down by weather, lack of fuel, a lost pilot, or mechanical problems. A communication system that we of today would regard as primitive saved the day for many a pilot. In 1922, and for years after, there were forty-eight U.S. Army Signal Corp stations in the Territory manned by 250 men and officers. They operated a mixture of telegraph and Morse-code-type radios in villages and remote stations. Telegraph wires were strung almost Territory-wide on the mainland. An underwater cable lay on the sea bottom between Southeastern Alaska towns and Seattle; a telegraph line connected Fairbanks and Seward.

It's winter, flying from Fairbanks, your plane noses over during a landing, and your metal prop hits the ground and is bent and unusable. If it is wood it has shattered. You hike along a dog team trail that is brushed out and maintained by the Territory, and maybe you'll get a dog sled ride, maybe not. Usually within a few days you arrive at a village or roadhouse with a telegraph or radio station where you can send a message to Fairbanks.

If a prop can't be sent by plane, it will arrive via dog team. It might be a scheduled dog team mail sled, or a hired dog team. This might take weeks, but that's the way life was then.

By the late 1920s cabin planes had started to replace the inefficient and frigid cockpits. Some of these planes had reliable air-cooled engines. At the same time, high wing monoplanes started to replace the biplanes.

In the beginning, the various commercial aviation companies and their pilots and mechanics resembled a small, sometimes quarrelsome family. Pilots and mechanics commonly changed from one company to another. Pilots often flew the planes of competing companies and no one though it unusual. When a pilot and his plane went missing, other pilots, with few exceptions, and regardless of company loyalty, participated in the aerial search for him.

In November, 1929, world-famous pilot Ben Eielson and his mechanic, Earl Borland, flew from the tiny coastal Alaska village of Teller into a snowstorm that raged over the Bering Sea. They were bound for the ice-locked trading ship *Nanuk* in Siberia which held a cargo of valuable furs. They were to fly the furs to Fairbanks to be transshipped to the New York fur market.

Their airplane didn't arrive at the trading vessel, nor did it return to Teller.

The winter search for the missing plane became the climactic Alaskan aviation event of the 1920s that was followed in news accounts by millions around the world. Between fierce storms searching pilots clothed head-to-toes in fur flew through brutal deep cold in open-cockpit biplanes. Daylight hours were dim and brief, for the sun remained below or close to the horizon.

Searching pilots first had to cross sixty miles of the ice-choked Bering Sea between Alaska and Siberia. They then followed the wild and barren Asian coast 375 miles to *Nanuk*, the base from which they flew their searches. Every foot they flew posed major flight hazards.

Today, despite the advances in quality and dependability of airplanes, even modern pilots would consider such flying as extremely risky.

It was late January before the missing plane was found.

Alaska's aviation family of the 1920s included heroic and bold pilots. They *had* to be adventuresome to do the flying they did. Those early fliers well-deserved the praise and affection bestowed upon them.

In this volume I have portrayed a handful of these men, concentrating largely on a few who pioneered aviation in Alaska's vast and rugged Interior, and who also participated in the search for the lost Eielson and Borland.

Alaska's aviation industry was a struggling infant during the 1920s. Eventually it became a giant upon which today's rural and not so-rural Alaskans are totally dependent. What would modern Alaska be without airplanes and bush pilots?

— JIM REARDEN
Sprucewood
Homer, Alaska

Acknowledgments

Richard Wien, lifelong resident of Fairbanks, and a long time commercial pilot, got me involved in the stories told in this volume. Not only that, to accompany the stories, he generously supplied beautiful, historic and irreplaceable photos from his father Noel's collection. Further, he carefully read early drafts and guided me through many aviation pitfalls. He contributed technical aviation aspects, as well as memories of his father's stories of the pilots and the times covered herein. Many thanks, Richard. Somehow that seems an inadequate way to express my appreciation for all your favors.

My wife Audrey, son Michael Rearden, and daughter Mary Bookman carefully read early drafts of this tome. Audrey's sharp eyes spotted typos and fuzzy wording; Mike the same, plus he did on-line computer research to fill many gaps I was unable to fill; Mary rearranged my comas, periods, and sentences into more acceptable form. My deep thanks for a family that didn't hesitate to criticize.

Homer Dr. Paul Eneboe, the Rearden family physician for four decades, and an avid reader, found an abundance of material in an early draft that needed reorganizing. I asked him for comments because I knew he wouldn't hold back, and I was right. He convinced me to make major changes from my original approach to this story. Thanks, Paul, you were right.

My thanks also to LD. "Corky" Corkran, CEO of the Pioneer Air Museum at Fairbanks, who was most helpful when I arrived at his fine museum to take photos and notes for this book.

Thanks to the Bob Reeve family for permission to use the Harvey Goodale portraits of Noel Wien and Ed Young. Early Alaska pilot Bob Reeve commissioned Goodale to produce a wonderful set of about two dozen portraits of Alaska's early pilots, including these two.

Mary Carlson at the Hatton-Eielson Museum, in Hatton, North Dakota, provided photos and helpful information on Ben Eielson, plus permission to use the elegant portrait of Eielson that appears on the cover of this volume. Many thanks, Mary.

Terrence Cole, Professor of History at the University of Alaska Fairbanks, obligingly read an advance draft of the manuscript and generously provided the comments that appear on the back cover. Many thanks, Terrence.

To the *Fairbanks Daily News-Miner*, long the finest newspaper in Alaska, thanks again for allowing me to use in one of my books quotes from your pages.

My thanks too to Stan Cohen, publisher, who continues to publish my books. Of the seven publishers who have printed my books, Cohen's Pictorial Histories Publishing Company has been by far the most satisfying. This book is my eleventh with his company's imprint. Stan Cohen's word over the telephone is more dependable than a written contract with some publishers I could name.

Kitty Herrin, of Arrow Graphics, at Missoula, Montana, has expertly designed all eleven of the books I have written for Pictorial Histories Publishing. Kitty's skill at presenting my clumsy writings in gracefully designed books is to me a constant source of amazement. Thanks, Kitty, for the quality of your work.

— JIM REARDEN
Sprucewood
Homer, Alaska

Book One

The Beginnings

Carl Ben Eielson in his twenties in the type of dress affected by outdoorsmen at the time; "chokebore" pants with leather puttees, and a pocket watch on a chain. Early pilots and others in Alaska often wore laced leather knee boots.

1 Ben Eielson Arrives at Fairbanks

Born at Hatton, North Dakota, in 1897, Carl Ben Eielson arrived in Fairbanks in 1922 to teach English and science in the high school and to coach the basketball team.

As a boy he was fascinated with airplanes, and decided he wanted to be a pilot. His father, Ole, a Hatton businessman, was opposed to the idea. "Too dangerous," he said.

Second Lieutenant Carl Ben Eielson when assigned to the Aviation Section of the U.S. Signal Corps. He was honorably discharged in 1919, and remained in the Signal Officers Reserve Corps.

Ben graduated from the Hatton high school and entered the University of North Dakota at Grand Forks, where he sang in the glee club, played cornet in the band, and joined the debating club. He transferred to the University of Wisconsin briefly, and in January, 1917, despite concerns of his father, he enlisted in the U.S. Signal Corps, Aviation Section, at Fort Omaha, Nebraska. In June, 1918, he transferred to the School of Military Aeronautics, University of California, Berkeley for eight weeks of ground school prior to flight training. He was then sent to Mather Field, near Sacramento, where he learned to fly. He was commissioned second lieutenant in the Aviation Section of the U.S. Signal Corps, and had orders to sail for France. The war ended, and his orders to France were cancelled.

He remained at Mather Field, and in March, 1919, he was honorably discharged as a Second Lieutenant, Signal Officers Reserve Corps. He chose to remain in the reserve corps.

The Hatton Aero Club

Back home in Hatton, Ben worked in his father's store, joined the American Legion, and talked aviation to anyone who would listen. His enthusiasm resulted in formation of the Hatton Aero Club during the winter 1919–20, which purchased a military surplus Model J1 Jenny for $2,485.

With this airplane Ben barnstormed in North Dakota that summer and flew to small town fairs for exhibition flights, which included acrobatics (aerobatics today). Eielson was reputedly an adroit stunt pilot.

That fall at Climax, Minnesota, while taking off from a muddy field, he wrapped one of the wings of the Jenny around a telephone pole. The plane dropped to the ground, one wing and the landing gear destroyed. Eielson was unhurt. The wrecked plane was hauled back to Hatton and the Hatton Aero Club was dissolved.

Ben rebuilt the damaged wing and landing gear, installed a new engine, and brought the Jenny back to flying status. He flew it to Grand Forks and re-entered the University of North Dakota. On weekends he barnstormed with the Jenny.

He graduated from the University with a Bachelor of Arts degree in June, 1921. That summer, two other WWI-trained pilots, Charles W. "Speed" Holman (who became a famed racing pilot later in life; Holman Field, St. Paul, Minnesota, honors his name), and Frank Talcott, joined Ben in barnstorming and aerial stunting exhibitions with the Jenny.

At Fairbanks

Ben sold the Jenny at the end of summer, 1921, and signed up for postgraduate law courses at Georgetown University, in Washington, D.C. To help pay his way, he worked on the Capitol Police Force as a guard in the U.S. House Office Building. There he met Dan Sutherland, Alaska's voteless delegate to Congress. He spent many hours visiting with Sutherland, mostly talking aviation with him. From Sutherland, Ben learned of a job opening for a high school teacher at Fairbanks in the fall of 1922. He applied for the job and was hired.

The Fairbanks he arrived at had a population of 1,155, and was the center of a gold mining district. When cold weather arrived in late October and November, miners from surrounding areas moved to town for the winter. The town was also the center for gathering raw furs, a secondary but important industry across Alaska. Homes were mostly log cabins, although frame buildings dominated the business district. There were seven hotels, eight restaurants, four dance halls. Electricity provided lighting for the town. Water was delivered by horse-drawn wagon in summer, and sled in winter.

Outhouses were common. Streets were unpaved, and there were more dog teams than cars. Winter sled trails to villages, mines, and trapping areas spiderwebbed from Fairbanks. In summer, a few cars traveled between coastal Valdez and Fairbanks on the Richardson Highway, an upgraded wagon trail that penetrated the great Alaska Range. There were 206 autos in private ownership in Fairbanks. The speed limit on the "highways" was twenty-five miles an hour. In winter, horse-drawn double-ender sleds traveled between Fairbanks and coastal Valdez. Roadhouses, roughly thirty miles apart, provided food and overnight accommodations. However, the Alaska Railroad, the northernmost railroad in North America, with 470 miles of rail from coastal Seward to Fairbanks, was completed in 1923, all but ending traffic to and from Valdez.

Living costs at Fairbanks ran somewhat higher than those in the states. Alex Simson's Department Store, opposite the Nordale Hotel on Second Street, advertised in the *Fairbanks Daily News-Miner* blue chambray work shirts for seventy-five cents, heavy pure wool socks for twenty-five cents. A suit of Medlicott wool underwear cost six dollars fifty cents. The Cut Price Store (which advertised, "Highest prices paid for raw furs"), sold heavy wool pants for five dollars; fine dress pants were seven dollars fifty cents. Stag wool shirts were six dollars fifty cents. Fairbanks merchants didn't accept coins smaller than twenty-five cents.

Eielson's Norwegian heritage proclaimed itself with his blue eyes and blond hair, already thinning at 25 when he arrived at Fairbanks. He stood a sturdy 5 feet 10 inches, and weighed 165 pounds. He was one of three teachers in the Fairbanks two-story, red, frame-built school, which, that fall had forty-eight students. Eielson was friendly, easy-to-meet, pleasant. He quickly made many friends in this tiny frontier town.

His students quickly learned if they could get him talking about airplanes, or aviation in general, he might take up a full hour period on the subject. Aviation, and the future of it, dominated his thoughts. Though he was new to the Territory, he already dreamed of a future when airplanes would provide passenger and freight service throughout Alaska. He even envisioned mail and passenger flights across Alaska to Siberia, and beyond to Europe.

His ideas were far ahead of the abilities of aircraft of the time; there were no airports as such needed for their support.

Fairbanks' First Airplanes

The first airplane ever at Fairbanks was a Gage-Martin biplane powered by an eight-cylinder Hall-Scott motor. It was owned and flown by its designer, James Martin, who was accompanied to Fairbanks by his aviatrix wife, Lily.

To transport their airplane to Fairbanks its wings were removed and crated. The Martins and their airplane traveled by ship from Seattle to Skagway. From there they rode the White Pass Railroad to Whitehorse. Next, the Martins and their plane went by river steamer down the Yukon to Tanana, and up the Tanana River to Chena, and up Chena Slough to Fairbanks.

The little airplane flew at about 45 mph, and between July 3 and 5, 1913, Martin made five flights with it from Fairbanks' edge-of-town Exposition Park. The longest flight lasted fourteen minutes.

Next, on August 19, 1920, the four De Havilland DH-4B biplanes of the U.S. Army's Black Wolf Squadron arrived in Fairbanks on their highly-publicized month-long New York to Nome flight. They were greeted by an enthusiastic crowd.

The squadron left for Nome the next day. After reaching Nome, they turned back, and again stopped briefly in Fairbanks on their return flight to the states.

A Trial Airmail Flight

Eielson often visited with sourdoughs in the lobby of the Alaska Hotel, where he roomed. Some told outrageous stories, hoping the young cheechako (newcomer) would bite. Among Fairbanksans who became his friends was debonair W. F. "Wrongfont" Thompson, editor of the *Fairbanks Daily News-Miner,* with whom Ben frequently visited. They talked mostly about aviation.

During winters, most mail in interior Alaska was hauled by dog team, and had been since the 1800s. It was slow and expensive. Eielson thought Alaska's mail could easily be flown, although no one had flown a plane through the deep cold of an interior winter. Somehow he managed to receive permission from the Post Office Department to make a trial airmail flight from Fairbanks to fifty-eight-mile-distant (by rail) Nenana.

With a borrowed airplane, on February 21, 1923, with the temperature at +5 F., with no wind, he flew 500 pounds of mail and express packages from Fairbanks to Nenana. The flight was recorded in the *Congressional Record.* References don't reveal which airplane he flew, or from whom he borrowed it.

The Farthest-North Airplane Company

Ben, with *News-Miner* editor W. F. Thompson and banker Dick Wood, formed the Farthest-North Airplane Company. They put together enough money to buy a military surplus OX-5-powered Jenny. Wood contributed most of the $750 price of the plane.

Eielson planned to fly the plane commercially—a first in interior Alaska.

The two crates that held the Jenny arrived at Fairbanks July 1, 1923. Eielson and Ira Farnsworth, "the best mechanic in town," worked at assembling it, with help from Earl Borland, a talented Alaska Road Commission mechanic. They carefully followed the thirty-five pages of directions that arrived with the plane.[1]

---

1. The book *Modern Aircraft*, by Major Victor W. Page, Air Corps Reserve, U.S.A.,1930, by The Norman W. Henley Publishing Company, New York, in my library, printed thirty-five pages of directions for uncrating, setting up, and aligning a Curtiss JN-4 biplane. —JR.

Eielson in the front cockpit of the Ox-5 Jenny NC47358, at Fairbanks. This airplane was purchased for $750 by the Farthest-North Airplane Company, which was formed by Eielson, W. F. Thompson, and Dick Wood. Passenger in the rear cockpit is Mrs. Ladessa Nordale, wearing Ben's flying helmet and goggles. Circa 1923.

THE JENNY FLIES

Finally, all parts were assembled and adjusted, with oil in the engine, and gas in the tank. On July 4, always a day of celebration in Fairbanks, Ben was billed as "The Greatest Living Flier, the Aerial Daredevil." The plane was rolled to a spot at the Exposition Park/ball park/race track[2] that gave Ben the needed room to take off. He climbed into the rear cockpit, put on his leather helmet, and pulled his goggles into position.

Farnsworth called, "Switch off" (the OX-5 had but one magneto) and Ben repeated. Farnsworth pulled the wooden prop through a few times, turning the engine over and priming the carburetor.

"Switch on," he called, and Ben turned the magneto switch to on.

The next time Farnsworth pulled the prop through, the OX-5

2. Exposition Park, at the edge of town, was used as a race track for cars, and for baseball. Eventually it became a landing runway for planes and named Weeks Field in honor of Secretary of War, John Weeks.

Adjacent land, owned by Paul Rickerts, was called Rickerts Field and used for a time as an airplane landing field. It was 1,500 feet long, as compared to the early Weeks Field of 900 feet. The first hangar built at Fairbanks was on Rickerts Field, in 1924, and was used by the DH-4 mail plane flown by Eielson.

engine sputtered a few times and stopped. He again pulled the prop through with the switch off. Next time, with the switch on, the engine started and the prop spun while the engine continued to run.

A huge crowd (for Fairbanks) had gathered. Ben allowed the engine to warm. Finally he advanced to full throttle. The OX-5 roared, and the Jenny bounced as it gained speed across the uneven ground and climbed into the Fairbanks sky.

For the next half hour he flew loops, spins, aileron rolls, flew upside down briefly, climbed, stalled, swooped near the ground and climbed noisily, above excited Fairbanksans.

When he landed, the day was still early. He took off and flew toward the fifty-mile-distant riverbank village of Nenana, where he had promised to give a flying demonstration. He followed the railroad tracks that ran between the two towns. His passenger was banker Dick Wood, who, despite prohibition, reportedly settled his nerves with a few shots of white mule before climbing aboard. Flying, after all, was a scary business.

Ben cut across bends in the railroad tracks, lost sight of the tracks, and circled, trying to find them again. Becoming lost was one of his weaknesses as a pilot; it happened with discouraging frequency. He circled for more than an hour, searching, and finally found the tiny riverbank village of Nenana. For $500 from the local citizenry, he performed loops, tailspins, aileron rolls, power dives, and other acrobatics.

He was called "The Flying Professor," and "Lieutenant Eielson," and various superlatives in the News-Miner's report on his day of flying.

President Warren Harding arrived at Fairbanks after driving a golden spike on July 16, 1923, just north of the new Tanana River bridge, symbolizing completion of the Alaska Railroad. Later, on a day at Fairbanks when the temperature reached 94 degrees, Eielson flew his Jenny in a spectacular series of stunts for the President and his party.

Commercial Flights from Fairbanks

That summer of 1923, with the Jenny, Ben Eielson flew passengers for brief flights, and to and from various mines and nearby villages. He flew machine parts to mines. He flew sick people from villages to the Fairbanks hospital. He flew game meat, gold, furs, groceries, and other items. Fairbanksans and regional miners found that air service was quicker, often by days, and often much cheaper, than ground or water transportation in the roadless wilderness-like land surrounding Fairbanks.

Advertising in the *News-Miner*, he offered, "Flying lessons from the Farthest-North Airplane Company. Your choice of the long course, or the short course! Take a dip in the clouds! Prices to fit all pocketbooks."

He flew commercial flights to Circle, Brooks (Livengood), Tolovana, Tofty, Stewart Creek, all within a short distance of Fairbanks. He was limited to light loads that would fit into the front cockpit, and by the 150-mile distance he could fly from Fairbanks with the Jenny.

Ben flew the Jenny for 145 hours that summer. Most flights were reported in the *News-Miner*. By summer's end various villages and mines surrounding Fairbanks had prepared runways where an airplane could land. Residents were beginning to understand the advantages of air travel.

Airmail for Alaska?

When cold weather came, the Jenny was stored for winter. While flying from Fairbanks that summer, Eielson had written to the Post Office Department in Washington proposing a contract for flying mail from Fairbanks to 300-mile-distant McGrath. In winter, mail between the two places was hauled by dog team, taking up to thirty days. Eielson pointed out that it was a simple flight, easily accomplished in a few hours. The Post Office Department responded with lukewarm bureaucratic double talk.

After all, in winter, mail had been carried by dog teams in Alaska since before the turn of the century. The routes were established, and a considerable business surrounded dog team mail carriers, including mail carrier contracts, sled builders, the sale of tons of dried salmon for dog food. Dog team trails ran all directions in Interior Alaska, with roadhouses spaced roughly every thirty miles where travelers, most of whom used dog teams, could find food, overnight lodging, and facilities for dog care. In addition, there were government-maintained relief cabins on long hauls between roadhouses. Mail team drivers were popular, for with teams of eighteen, twenty, or more dogs, they broke trail in new snow. Other users of the trails often waited for a mail team to go by so they could follow on the broken trail.

Mail dog team drivers annually bid on mail routes. Those awarded contracts for the coming winter were listed in the June 2, 1926, *News Miner.* A sampling (value of contract not listed):

Bethel to McGrath—Wallace Langley
Bethel to Quinhagak—Charles E. Jacobson
Bettles to Wiseman—Sam Dubin
Candle to Keewalik—Herbert Greenberg
Chatanika to Circle—Henry Robson
Chicken to St. Timothy—R. C. Mitchell
Circle via Fort Yukon to Beaver—Northern Commercial Company
Circle to Miller House—John Palm

A dog team of about twenty animals used by a mail team driver. Winter mail was hauled by dog teams in mainland Alaska from the 1800s until airplanes took over mostly by the late 1930s. This photo was probably taken at Nenana. Commonly, such a team pulled two heavily loaded sleds hooked together. Once airplanes became established, winter mail could be flown more economically than hauled by dog teams.

Within a few years the airplane forced many of the enterprises involved with dog teams out of business, but not without a struggle. As airplanes challenged, a notice posted outside one roadhouse read, "Drunks, Indians, and Airplane Pilots not Welcome Here." The airplane changed life in Alaska in many ways, and much of the change took place within ten or fifteen years. Those whose livelihoods were displaced generally didn't like pilots.

That fall of 1923 Eielson traveled to Washington D. C. attempting to obtain a contract to fly mail from Fairbanks to McGrath. He visited Post Office officials and pitched his proposal. He reminded authorities of his February airmail flight to Nenana. He was a handsome young man, full of enthusiasm, and persuasive, and he had flown as an Army pilot, which was all a plus. Alaska was a complete unknown to the bureaucrats he approached. They probably regarded Eielson as a strange bird.

McGrath lies at the confluence of the Takotna and Kuskokwim Rivers, and is a center for gold mining, trapping, and freighting. It is a transfer point for freight bound for the upper Kuskokwim River; here larger river boats and tugs with barges transfer freight to smaller vessels better adapted to the upper reaches of the winding river.

Flying the mail was relatively new in the United States, although it was common in Europe.[3]

An Airmail Contract

Dan Sutherland, Alaska's Delegate to Congress, had been promoting air mail service for the Territory for three years. Eielson's pitch and Sutherland's pressure worked, for in December, 1923, the Postal Service announced a contract would be let to the Farthest-North Airplane Company (Eielson, W.F. Thompson, and Dick Wood) for an experimental aeroplane service for the 230 miles between Nenana, on the Alaska Railroad, and McGrath, on the Kuskokwim River.

3. In 1918, a trial airmail run was made between New York and Washington D.C., using Army pilots. The Post Office Department followed up using Standard biplanes, and later, with modified War Surplus DeHavilland DH-4 biplanes. By 1920, daylight mail flights extended from New York to San Francisco; railroads took the mail over at night. Next came night flights on which mail plane pilots followed revolving beacons. By mid-1924 revolving beacons flashed into the night sky from New York to San Francisco, with lighted emergency landing runways every thirty miles or so. Night and day airmail flights, weather permitting, were well established.

The route crossed the Kantishna gold mining and fur trapping country, with stops at Ophir, Takotna, Iditarod, and Flat City. Winter dog team mail service was to continue during the airmail flights. Ben Eielson was to pilot a DeHavilland DH-4BM biplane loaner[4] from the Postal Service, with a 440-hp Liberty engine. It included special equipment for cold weather operation.

Skis and a spare motor were included with the crated DH-4 shipped to Fairbanks. It arrived via the Alaska Railroad on January 23, 1924. The three huge crates were hauled from the rail depot by a horse-drawn sled to the Northern Commercial Company machine shop for assembly by Eielson and mechanic Frank McCafferty.

Assembled except for wings and empennage (tail feathers), on skis, with a dog sled under the tail, the fuselage was pulled by a horse from the machine shop to an unfinished hangar at Rickert's field. Here the wings, tail ski, and empennage were attached.

Eielson was to make ten round-trip trial flights between Nenana and McGrath. Payment was to be two dollars a mile for the first five round trips, and a dollar fifty for the remaining five trips. That was less than half what the dog team musher with the mail contract received.

Instead of starting his flights at Nenana, Eielson based himself at Fairbanks with its better machine shop and other facilities. This added fifty miles to the one-way distance to McGrath.

First Mail Flight to McGrath

It was -5 F. on February 21, 1924, at Week's Field when, at 8:45 a.m, Eielson, with 164 pounds of mail in the enclosed front cockpit, gave full throttle to the Liberty engine of the DeHavilland, and roared down the 900-feet-long runway to climb into the sky on his first mail flight to McGrath. The big wood prop created such

4. About 300 DH-4 light bomber and observation planes were sent overseas during World War I, but none were used in combat. At war's end there were many surplus DH-4s,and the Postal Service started using them in the States for hauling mail. Before such use each plane had an engine, propeller and instrument overhaul, had the cockpits realigned, with the front cockpit with a hinged cover. The airmail pilot's name was printed on the fuselage.

Army pilots referred to the DH-4s as "flaming coffins." The fuel tank was placed low between engine and cockpit. Air pressure in the tank was used to force fuel to the carburetor. A punctured fuel tank or fuel line sprayed fuel. When that happened, a fire often resulted.

Ignition for the 440-hp Liberty engine was provided by a hot battery. A dead battery meant a stopped engine, and that meant the airplane had to immediately land.

Ben Eielson in the cockpit of the DeHavilland D4 in which he flew the first mail flight from Fairbanks to McGrath on February 21, 1924. Eielson was the first to fly scheduled mail runs in Alaska in a trial effort by the Post Office Department.

a blizzard of snow that the airplane wasn't visible to onlookers until it climbed free of the ground.

The compass of Ben's DeHavilland was forty degrees off. The airspeed indicator was inoperative, and the tachometer worked intermittently. He ignored these problems and flew on.

He was heavily bundled in layers of wool and fur, and carried emergency gear of snowshoes, a mountain sheepskin sleeping bag, ten days provisions, axe, gun, and tools for working on the plane.

Ben followed the Tanana River to Nenana and from there followed the dog team trail southwest. He peered down trying to spot the roadhouses and their smoke every thirty miles or so. Reaching the winding Kuskokwim River, he followed it to McGrath and landed on the Takotna River where it poured into the Kuskokwim. The flight lasted two hours and fifty minutes.

A dog team hauled the mail to the McGrath post office. Sixty pounds of mail was returned to the plane in the dog sled for the return trip. The plane was refueled and oil added to the engine.

Eielson had planned a quick turn-around, expecting to land at Fairbanks by dusk at around 5 p.m., but McGrath locals insisted on a celebratory banquet, which delayed his departure until 2:35 p.m.

As dark fell he was about half way to Fairbanks and above nine-mile-long Lake Minchumina. He continued with the same heading. When he thought he should be in the vicinity of Nenana he could see no lights. He circled, peering for lights. He homed on a light at a cabin, and when it was beneath him he realized it was in a remote area.

For an hour he was lost in the dark. Dense clouds concealed the stars; it was black dark. Only the snow-covered land gave Ben some idea of the lay of the land below. He grimly flew on, carefully searching for lights. A glimmer anywhere would do at this stage. He was going to have to land soon. Gas was running low.

He came to a large river, followed it, saw a flare in the distance, flew to it, and discovered with great relief it was a bonfire for his benefit in front of the Rickert's Field hangar.

He was unable to see the edge of the landing field, for it was 6:45 p. m. and full dark as he glided for a landing. Missing the edge of the field, he struck a tree and one of the plane's skis broke off. The plane nosed over when it hit the ground, and as the waiting crowd watched, the propeller broke.

Fairbanksans proudly presented Eielson with a gold watch, inscribed, "C.B. Eielson—Pioneer Alaska Trail Blazer—Fairbanks to McGrath—February 21, 1924." Fastened to the watch was a gold chain that included a gold nugget; a diamond-studded knife was attached to one end of the chain.

He was made an honorary member of the Fairbanks Igloo of Pioneers. He gave a talk about aviation and his airmail contract to the faculty and students at the Alaska Agricultural College and School of Mines (today the University of Alaska Fairbanks).

More Mail Flights to McGrath

The compass in the DeHavilland was repaired, and the airspeed indicator fixed for the second mail flight on March 1, on which he carried 252 pounds of McGrath-bound mail. It went off without a hitch.

The third flight, on March 12, also went well. He hauled little mail, but he took thirty-five pounds of wire for the government radio station at Takotna, twenty miles from McGrath, and landed there to deliver it.

In powdery snow the narrow tail ski sent with the DeHavilland created drag in the snow on takeoff. It was replaced by a six-inch-wide, shorter ski which worked better.

Eielson flew his fourth mail flight on March 26, and had a minor problem on landing on the Takotna river when he ran into an overflow which had been covered by snow. The ice six inches beneath was solid, and the only damage was a cracked ski, which was temporarily repaired for the return flight.

The fifth trip, on April 9, with 300 pounds of mail, on a blue sky day, went without a hitch.

On his sixth trip, April 23, he landed at Nenana to take on 365 pounds of mail that had arrived there by train the previous day. The last dog team over the route had left for McGrath April 5, and surface travel had ended because of breakup. The first upriver boat on the Kuskokwim River to McGrath wasn't expected until about June 20. Eielson pointed out that breakup didn't stop his airplane from flying.

At McGrath, William "Hosie" Hummel, the "High-powered Swede,"[5] extremely ill, had been hauled by dog team from Takotna. There was no doctor within hundreds of miles. Ben loaded him into the DeHavilland with the mail and flew him to Fairbanks for medical care. He charged nothing.

After landing at Fairbanks, Ben attempted a fast turn on the ground and broke the pedestal on one ski as well as the airplane's wood propeller. In addition, the radiator was damaged when the nose of the plane hit the ground. Repairs were made.

As Ben helped Hosie out of the cockeyed plane, the ill man reportedly said, "Yeezus, Ben! You always land like dat?"

Snow was gone and the skis were replaced by wheels for the seventh flight on May 7, which went well.

Flight number eight, on May 28, ended after Ben landed the DeHavilland on return to Fairbanks. He taxied the DeHavilland into a boggy spot (later spoken of as "Eielson's soft spot") near the center of the airfield, where the wheels sank deeply in mud. The plane flipped onto its back. Propeller, rudder and two wing struts broke.

In the mail cockpit, passenger Charles Nystrom, of McGrath, who was making a hurried trip to the Mayo Brothers Clinic in Rochester, Minnesota, for an unknown health problem, hung by his safety belt. Ben released it and his passenger promptly landed on his head. Except for a goose egg, Nystrom was all right.

Ben broke the DeHavilland three times on eight flights. In fairness, it must be said that the big, awkward airplane was not well suited to the extreme winter conditions, or to the available

5. Many of the colorful characters of the time had equally colorful nicknames, a custom that, sadly, has mostly disappeared in Alaska.

landing places.

All the spare parts needed for repairs that had been sent with the DeHavilland had been used. Local mechanics couldn't repair it this time without sending Outside for parts.

The Post Office Department was notified of the accident, and refused to send or finance replacement wing struts. The Assistant Postmaster General wrote, "Your experiment has been successful to a marked degree...(but) there are many things which must be done before we can continue on a permanent basis our use of airplanes in mail-carrying in Alaska."

So ended the first real attempt to haul mail by airplane in Alaska.

After his last mail flight, for a time that spring and early summer of 1924, Eielson flew the Jenny briefly at Fairbanks for Jimmy Rodebaugh, an Alaska Railroad conductor who was gearing up to get into the airplane business. On September 19, 1924, for a $200 charter, his passenger was miner Jack Tobin, whose destination was near Copper Mountain in Mount McKinley National Park. Copper Mountain was later renamed Mount Eielson in honor of Ben, who was the first to land an airplane near it.

In his first two years (1922–24) at Fairbanks, Ben Eielson became Alaska's first airmail pilot, as well as the first pilot ever to be based at Fairbanks with a commercial flying operation.

That June, Eielson traveled to the states. On the way, while in Anchorage, he was attracted to a newly arrived Standard biplane being flown by Noel Wien from a new airstrip. The two pilots met and talked, and Eielson encouraged Wien, new to Alaska, to pursue flying in Alaska. There was a strong mutual attraction. They were both products of the rural Midwest. Both were aviation enthusiasts. Both would become famous for their flying exploits. Eventually they became fast friends.

Ben again made the rounds of officials in Washington D.C. in attempting to persuade the government to establish air mail in Alaska. It was like catching water in a sieve; most officials had unchangeable views on Alaska. To these bureaucrats, the old Eskimo and snow igloo fable and impossibly cold winters were established facts despite Eielson's assuring them otherwise, and his proven flying experience in the Territory.

He returned home to Hatton, briefly returned to Georgetown University law school, then dropped out to join the Army Air

Service to participate in a cold weather flying study which was finished by the end of 1924, along with his contract with the Army.

Back at Hatton, Ben became a bond salesman, which bored him. In early 1926, a telegram from explorer George Hubert Wilkins rescued him. Wilkins was looking for a pilot to fly for him on an arctic expedition. Vilhjalmur Stefansson, famous as an arctic explorer, who had followed newspaper accounts of Ben's airmail flights, had told Wilkins that Ben was the American pilot with the most experience in arctic flying.

Ben met Wilkins in New York. They were both adventure-minded and both were pilots and aviation enthusiasts. They enjoyed each other's company, and quickly came to agreement. For a nominal salary, Ben agreed to accompany Wilkins as a pilot on the proposed arctic expedition.

Ben Eielson's life was about to change; as an Alaskan pilot who first flew airmail in the Territory he had gained moderate fame; as a pilot for Hubert Wilkins, he was to become internationally famous.

2
Noel Wien Arrives at Fairbanks

Noel Wien, born June 8, 1899, was one of five children born to immigrant parents—his father was from Norway, his mother from Sweden. He grew up in a log cabin on the family homestead at Cook, in northeastern Minnesota where the Wiens lived largely from the land.

Fascinated by mechanical things, cars were his first love. He was 10 when he first had a brief turn at steering a car—a four-cylinder Elcar. His father bought a used Model T Ford when he was 17, but, for a time, Noel was considered too young to be allowed to drive it.

He was a teenager during World War I, when he read about Mannock, Rickenbacker, Fonck, Guynemer, Nungesser, and other famous military fighter plane pilots on the Western Front. He soon became knowledgeable about the Curtiss JN4 "Jenny," the main training plane for Americans at the time, as well as WWI fighter planes, the Spad XIII, the British S.E.5, the Sopwith Pup, the Nieuport 17 and others. He learned about airplane motors, (now called "engines") used in various planes - the Hispano-Suiza (Hisso), Gnome, LeRhone, Liberty, and OX-5. In later years he wryly commented that it would probably have been better had he paid as much attention to his schooling as he had to cars and airplanes. His school attendance ended when he was 18, after he had completed eighth grade for the second time.

While still a teenager, he decided he wanted to spend his life flying, although he had never been near an airplane.

The summer he was 17, while continuing to live at home, he raked rocks on county roads at fifteen cents an hour; $1.50 for a ten-hour day. Next, for $2.50 a day he drove a two-ton GMC dump truck, hauling, instead of raking, rocks.

On his 21st birthday Noel bought a 1920 Overland touring car with the nearly $800 he had saved from his two jobs. He lost his truck-driving job and went to Duluth and worked at a harness factory riveting buckles on and oiling harness for $1.50 a day. Horses were still an important means of transportation and farming power. He lived at the YMCA, and ate one meal a day.

He sold the Overland and went to Minneapolis and Saint Paul and used the money to sign up at the William Hood Dunwoodie Institute, hoping to learn about airplanes. There was no airplane course, so he signed up for the auto mechanics course.

Finding an Airport

He soon discovered the nearby Curtiss Northwest Airplane Company's flying school which had a landing area that resembled a forty-acre hayfield. After that his life centered on airplanes and landing fields; he never returned to the Dunwoodie Institute.

In May, 1921, at the flying school, he finally laid his hands on an airplane and met Major Ray S. Miller, a well-known Minnesota pilot.

"I can teach you to fly in eight hours of flying time. It will cost you $40 an hour," Miller told Noel. "A demonstration hop will cost you $10."

The ten minute demonstration flight on May 6, 1921, in a Curtiss JN4, Jenny, extended into twenty minutes. Miller tested his prospective student by flying loops, spins, stalls, and wingovers. If he survived with a grin, the instructor believed, he might make a pilot.

Noel loved it. He had dreamed of flying for years, and now that he was in the air he could hardly believe it. He peered at the world below, seeing miniature buildings and farm fields and fences as through the wrong end of a telescope. He watched insect-size cars crawl along dusty roads. The dizzying aerial maneuvers Miller flew didn't make him airsick; they thrilled him.

When they landed, he advanced $40 to Miller for the first four fifteen-minute instructional hops.

Flying a Jenny

Miller taught him to fly in an OX-5-powered Jenny, a biplane (two wings) which had no airspeed indicator; he had to judge its speed by the pitch of the air passing across the many wires that held the airplane and its two wings together. There was no turn and bank indicator. A coordinated turn depended on the sensitivity in a pilot's rear; flying was largely a seat-of-the-pants experience; wind on one side of an open-cockpit-occupied pilot's face could hint at an un-coordinated turn or skid.

The water-cooled ninety-horsepower OX-5 engine of the Jenny required ten minutes to pull the airplane to 2,000 feet. Its payload was 490 pounds, which included the pilot, a passenger, gasoline, and oil. It was far from the docile airplanes of the mid-20th century; it could fall into a stall and spin with little warning. There was little glide to it; it dropped like a rock without power. He learned how, from altitude, to identify a corn field, other grain fields, or a potato field for emergency landings, which he could expect at any time. The OX-5 was not as dependable as later engines.

During his third hour of dual instruction, he landed the Jenny without the instructor touching the controls. By his fifth hour, he repeatedly took off and landed without the instructor's help.

An occasional individual is born to be a pilot. It requires perfect coordination, an understanding of the controls and their use, and an appreciation that the airplane moves in three dimensions. Noel Wien was such a person. He was ready to fly solo after eight hours of instruction. At the time, a newly soloing pilot had to guarantee to replace a broken airplane before being allowed to take it up. Noel couldn't afford to pay for an airplane, or the necessary bond that would allow him to solo. After eight hours of instructions, he left the Curtiss Northwest Company without soloing.

Flying was everything Noel had dreamed it could be. He was hooked. The sense of freedom that comes from soaring high, seeing the earth as the birds do, the ability to dive, turn, to skim near clouds, is almost indescribable. There is an intense delight in flying in an open-cockpit plane like the Jenny, the type that dominated the early years of aviation. He could look straight up into the boundless sky, or straight down to the earth; his vision was wide open. The airplane seemed to be almost a part of his body. He was not enclosed and bound, as in a cabin plane; there was a simple and wonderful sense of freedom.

When those who flew in cockpit type planes for years converted to cabin planes, most complained, "I can't see like I could from a cockpit."

They got used to it. Being warm was their reward.

First Flying Job

Shortly after completing his dual instructions, Noel looked for a job of any kind that involved airplanes. His instructor introduced him to E. W. Morrill, a former Navy pilot who owned a World War I surplus Standard biplane which he planned to use for barnstorming. Noel agreed to work as his helper. He would build time as a pilot by helping to fly the airplane cross-country between barnstorming gigs. He would also help maintain the airplane, collect passenger's money, or whatever came along. In exchange he was to receive food and lodging; no dollars.

Early during their barnstorming tour they arrived at a small Minnesota town over which Morrill performed the usual noisy barnstormer's gyrations with the plane to attract the attention of potential passengers. He then headed for a small field from which he planned to operate. To Noel, in the front cockpit, the field looked too small. Since Morrill was an experienced pilot, at first he wasn't concerned. But when Morrill tried to land downwind and with a slight crosswind, he took notice.

Twice Morrill tried to land, having to pull up at the last moment each time when it became obvious the plane wouldn't stop before running into a stand of corn. As Morrill tried the same approach for the third time, it was apparent to Noel that a landing could result in disaster. He seized the control stick, pushed the throttle wide open, and lifted the plane clear. Once at a safe altitude, he glanced back at Morrill, who, to Noel's surprise, raised both hands, indicating he had relinquished the controls. Noel was now in charge.

He circled, flew an upwind approach over tall trees at the edge of the field, and dropped the plane into the tiny field with a perfect three-point landing. The Jenny stopped a few feet from the corn. Noel cut the engine and looked back at Morrill. He had transgressed by seizing control from an experienced pilot, and the owner of the airplane at that. He expected a strong rebuke, perhaps he would be fired.

Instead, Morrill said, "Good work. I couldn't tell which way the wind was blowing." Wien had sensed the wind direction, which told him the proper direction from which approach to the field. In essence, that was Noel's solo flight; he had command of an airplane at a critical time.

Morrill further acknowledged Noel's skill by telling him to fly the plane out of the little field. "We won't barnstorm from here—the field is too small," he said.

Noel barnstormed with Morrill into August, 1921, by which time he had logged about seventy hours as a pilot.

Barnstorming and a Flying Circus

For the next three years he barnstormed with various partners. He also worked for a flying circus that operated from Minnesota to California, and on to Texas. He became skilled in flying a loop with a wing-walker standing on the top wing; he flew parachute jumpers, and did acrobatics during circus performances.

In later years as a pioneering Alaska bush pilot, Noel Wien's reputation was that of a safety-conscious, conservative pilot who commonly flew *around* rugged country and big timber to provide an extra margin of safety in event of a forced landing. Except in an emergency, he refused to challenge bad weather.

Alaskans who knew and admired the safe and careful Noel Wien would likely have been surprised to learn about the dust-'em-up and turn-'em-over kind of flying he had done in his early years as a pilot.

Alaska Bound

Wien's early barnstorming and other flying jobs never lasted more than a few months. He was at home at Cook, Minnesota in May, 1924, when he was hired by James S. "Jimmy" Rodebaugh, Senior Conductor on the Alaska Railroad who had accumulated a stake trading furs along the rail belt. Rodebaugh thought airplanes could be useful in Alaska. He had bought two crated J-1 Standard biplanes from airplane dealer Marvin Northrup at Robbinsdale, Minnesota (not related to the airplane builder Northrop).

Northrup improved each of the Standards by removing the

Mid-summer, 1924, at Fairbanks and the two Hisso Standard J 1s owned and shipped to Alaska by Jimmy Rodebaught (on left). During the summer of 1924 and 1925 these were the only two airplanes being flown in Alaska. Pilot Noel Wien is on the right wearing laced leather knee boots and breeches common to pilots of the time. Eddie Hudson, miner, in center.

bucket seat from the front cockpit and replacing it with a bench that allowed room for two passengers; he replaced the landing gear with the more rugged De Havilland DH-4 gear; the Hall-Scott original motor was removed and the motor mounts rebuilt to accommodate the more powerful Hispano-Suiza 150 hp motor. He removed the vertical radiator, replacing it with a nose radiator, giving the pilot better visibility. The airplanes were assembled, flight tested, and, still in their olive-brown military paint, except for their red, white, and blue tails, and the Army Air Service roundels on the wings, disassembled and crated, ready for shipment.

Rodebaugh asked Northrup to hire two pilots and a mechanic to accompany the planes to Flairbanks.

The two pilots Northrup sent were Noel Wien and Art Sampson.[1] The mechanic was Bill Yunker, who was also a pilot, and with whom Noel had once briefly barnstormed.

The steamer *Northwestern* docked at Seward, Alaska, in early

1. Art Sampson arrived at Fairbanks by ship and train with the first Standard, and flew from there briefly. He then waited arrival of Noel Wien so he could leave. "This is no place for an airplane. There's no place to land," he told Wien. He returned to Minnesota and for 25 years he was Head of the Aviation Department of the State School of Science at Wahpeton, North Dakota. He also organized the North Dakota Wing of the Civil Air Patrol. After retirement he operated a welding and repair shop until his death on June 3, 1962.

June, 1924, with 25-year-old Noel, now with more than 500 hours as a pilot, and a crated Standard aboard. Yunker, who had already been to Fairbanks with the other Standard, was there to meet him and the plane.

At Anchorage

The crated plane rode an Alaska Railroad flatcar to Anchorage, then a railroad town of 2,000, where Noel and Yunker assembled it and had "ANCHORAGE" in huge letters painted on its fuselage.

That June, in Anchorage, Noel flew more than 170 passengers on sixty-five flights with the Standard from a 2,000–foot runway volunteers had prepared just south of the town. He charged ten dollars for a fifteen-minute ride, and took in $1,700—a tidy sum in 1924.

Today, the runway he used is the Park Strip that parallels Ninth Street in down-town Anchorage.

The Standard, with Noel piloting, was the first airplane ever to fly passengers from Anchorage. On July 4th he flew loops, spins, and stalls for a large and appreciative Anchorage crowd.

Yunker installed a streamlined auxiliary fuel tank on the underside of the upper wing. It was built to his specifications by local resident Oscar S. Gill. This gave the plane sixty-five gallons of gasoline, with a range of about 400 miles, enough, Noel hoped, to allow him to safely reach Fairbanks.

While in Anchorage, Noel met Carl Ben Eielson, who was on his way out of Alaska. Within a few years Eielson was to make a name for himself as a pilot in Alaska and elsewhere. In time, he and Noel became friends.

"You'll like it up here, and you'll do well," Eielson predicted.

Flight to Fairbanks

On July 6, 1924, at 2:30 a.m., Wien, with Bill Yunker in the front cockpit, lifted the Standard from the Anchorage runway and flew to Fairbanks. It was the first flight ever between the two cities, and one of the dozens of first-flight records in the Territory made by Noel Wien during his early years of Alaska flying.

Noel was not one to seek fame. He wrote the following brief

June, 1925. Noel Wien landed on the first airport built in interior Alaska by Frank Leach, owner of Circle Hot Springs. "I went up on the nose and the prop stuck straight down into the ground, but did not break. We used the horses to pull the plane to harder ground, and take off was made without trouble. Frank Leach with a prying pole on left left, helped by Joe Mehern on right

account describing the historic flight in the August, 1956, *Wien Alaska Arctic Liner* monthly publication, which was aimed mostly at company employees.

Recounting the Early Days of Wien Airlines
By Noel Wien

The change has been great both in aviation and the city of Fairbanks since that memorable day, July 6, 1924, when, in a water-cooled Hisso-powered Standard J-1 open-cockpit biplane, Bill Yunker and I landed here after flying non-stop from Anchorage.

We flew up at night, thus taking advantage of the smoother air. The smoke was very thick for the last eighty miles and kept us guessing all the time. It was even difficult to follow the railroad tracks from Nenana on in.

[Author] *The flight wasn't as simple as Noel made it sound in the above recounting. Following the railroad was easy, as long as it was visible; pilots still follow railroads as a navigation aid, and Noel had often followed them in the states. However, on this flight there were no airports for a safe forced landing. Much of the land was rough and steep, impossible to consider for a landing. The Alaska Railroad map he used showed towns and stations, but the "towns" often consisted of a few small scattered buildings. Stations were commonly one small building. The railroad had been completed the previous year, and there was little development along it; Noel flew over mostly unsettled wilderness.*

After passing Mount McKinley (Alaskans call it Denali), North America's tallest peak, he left the foothills of the Alaska Range with relief when he reached relatively flat, green country. It wasn't farming country with pastures and cultivated fields he was accustomed to; instead, it was miles and miles of uninhabited tussocky tundra, interspersed with patches of spruce and birch trees. There were no obvious places to land safely.

Ahead of his airplane, the railroad tracks disappeared in forest fire smoke that extended to 10,000 feet. To keep the tracks in sight, he had to fly at 200 feet. Visibility grew worse, and smoke burned his eyes. He was soon forced to fly at 100 feet. He kept his eyes on the railroad, afraid to look elsewhere, fearing he might lose the tracks at a turn.

Noel Wien, on right, bound for the Kantishna area with mining engineer Ingram (left) and his secretary, Billie Hart, in September, 1924. Bad weather forced Wien to land on a 300-foot long bar on Bearpaw Creek twenty miles short of their destination. His passengers had to walk the rest of the way.

Three hours and forty minutes from Anchorage, Yunker pointed down to the University of Alaska's experimental farm. Four miles farther, Fairbanks itself was covered solid with smoke, and Noel was startled by the sight of the two 200-foot-high mid-town Northern Commercial Company's smoke stacks. He quickly banked away from them. Shortly, he saw and landed on the race track at the edge of town. Race tracks (for cars) were commonly used by barnstormers in the states, and experienced barnstormer Noel had found a home.

That race track eventually became Weeks Field, for many years Fairbanks' only airport. The town grew around it and it was clearly too small and inadequate for the use it was getting, as well as being in an awkward location, when it was finally closed in October, 1951.

PLUCKY AIRMEN BRING "ANCHORAGE" TO INTERIOR WITHOUT STOP; FAST TIME MADE OVER UNKNOWN COURSE, read the Fairbanks Daily News-Miner headline for the day, with the following somewhat misleading report:

> Arriving over Fairbanks at 7:16 o'clock this morning, the airplane ANCHORAGE, ship No. 2 of the Alaska Aerial Transportation Company, successfully completed the first non-stop flight between Anchorage and Fairbanks. The flight was made in the fast time of 3 hours and 45 minutes, and was accomplished without incident. Pilot Noel Wien was at the stick, with Mechanician William B. Yunker[2] as passenger.
>
> Wien and Yunker, after making a pretty landing on Weeks Field, stated that, although they followed the general course of the Alaska Railroad, at no time was the roadbed visible to them. They were able to discern the Carlson roadhouse at Cantwell, where a landing field was said to be ready for them, but they were unable to distinguish the field.
>
> Mount McKinley, rising to majestic height, was not visible until they were within a short distance of it. The lofty dome presented an inspiring pictured veiled in the low early morning mists surrounding it.

2. Bill Yunker, a mechanic, and also a pilot, remained at Fairbanks as Chief Mechanic for Alaska Aerial Transportation Company until September 9, 1924, when he left for his home in Rochester, Minnesota. He managed the Rochester airport, worked as a mechanic for American Airlines in Chicago, and later worked for North Central Airlines in Minnesota as a mechanic, where he became Engine Overhaul Superintendent. He died October 29, 1960.

Noel Wien (left), in 1925, with one of the Hisso-Standard biplanes which he flew commercially from Fairbanks in 1924, 1925, and 1926. His passenger here was gold miner Carl C. Dunlap who Noel flew from Fairbanks to Beaver on May 3, 1925.

Seventy-five years later, on July 6, 1999, with special permission from the Federal Aviation Administration, another biplane, a World War II Stearman with two open cockpits, took off from the identical location used by Noel, now a park in mid-Anchorage. It flew to Fairbanks to repeat and commemorate Noel's historic flight. At the controls were professional pilots Noel Merrill Wien and Richard A. Wien, the two sons of Noel and Ada Wien.

3
The Early Days of Wien Airlines

By Noel Wien

[Author] *During the mid-1950s Noel Wien wrote a series of "looking back" articles for the* Wien Arctic Liner, *an inhouse monthly publication for the airline. Selections from these first person remembrances, slightly edited, follow.*

There was intrigue about the stillness of the air, and the frontier atmosphere of Fairbanks, which made me like the North from the day I arrived. For two weeks after we landed [on July 6, 1924; "we", meaning Noel] we couldn't find our way cross-country due to the forest fire smoke, but when it cleared, we were busy. People in Fairbanks took to the air quickly. They were hardy, willing to gamble. Ben Eielson had made a number of flights that spring before I arrived [for Rodebaugh's Fairbanks Airplane Corporation). He had also started the Farthest-North Airplane Company the previous year, and had brought in an old reliable OX-5-powered Curtiss Jenny JN-4D open cockpit World War I training plane.

Due to the interest created by Eielson's pioneering, we had little trouble getting flying business to outlying mining camps. Livengood, sixty miles northwest of Fairbanks, a cluster of mostly log cabins surrounded by mines, was one of the best of the gold-producing camps.

During the first season in 1924, we made thirty-four flights to Livengood, and in the summer of 1925, forty three flights.

Broken Water Pump

All went smoothly until mid-summer 1925. We had purchased a supposedly major-overhauled plane from Lincoln, Nebraska, one of the Hisso-Standard build-up headquarters. The engine worked fine on the flight to Livengood, but on the return a sudden shower of water erupted from the engine's cooling system.

I knew that because of loss of water, the engine would soon become so hot it would stop. We were about half-way to Fairbanks, near Wickersham Dome. I spotted a shelf to one side of the Dome which seemed like the only possible chance of getting down without breaking up or going over on our back. We were cruising lower than the 2,500-foot shelf, so I had to use power to get up to it. The engine was steaming plenty when I reached a landing approach.

It was a fairly good landing place, and the airplane remained right side up without breaking anything. The problem was caused by a broken water pump; water stopped circulating, overheated, and boiled over.

The two passengers and I walked to Olnes, on the Chatanika River [fifteen miles], over miles of miserable tussocks. One of the passengers, an old sourdough, had no trouble walking. The other passenger, an insurance adjuster, had flown with me for both business and pleasure. He was my first tourist, and possibly the first flying tourist passenger in Alaska. He wore Oxford shoes and was about to give up before we arrived at the Chatanika River.

After one day of rest, I was ready to attempt another landing on Wickersham Dome. We had just hired A.A. Bennett, a new pilot, to help us fly one of the three ships we had [the two Standards and Eielson's Jenny]. He was new to Alaska, having just arrived from San Diego, California, and was skeptical of going in with me. Our second Hisso Standard was not in shape to fly at the time, but we had Ben Eielson's OX-5 Jenny in good flying condition, so in it off we flew to Wickersham Dome to retrieve the crippled Standard.

The day was nice and I made a good landing, but busted a tire on a protruding rock. Knowing the landing surface was rocky, we had with us an extra wheel and tire already assembled. After replacing the wheel on the Jenny, we replaced the broken water pump on the stranded Standard, and re-filled the radiator with water we had carried with us from Fairbanks.

Noel Wien worked for the Bennett-Rodebaugh company 1924-26. The Stinson Standard airplane seen here is the same model plane as the Stinson Detroiters brought to Alaska by Hubert Wilkins.

L to R. A.A. Bennett, Ed Young, Jimmy Rodebaugh, unidentified, pilot Matt Nimienen, and Leonard Seppala, one of Alaska's most famous dog mushers. The woman in the sled is probably Sigrid, Seppala's daughter. Driver of dog team not identified. ED YOUNG COLLECTION

I gave Bennett his choice of the two ships in which to make the attempt on take-off. Having more experience in the OX-5 Jenny, he selected it for the return flight.

The shelf on the dome was close to 2,500 feet above sea level. The Jenny was quite heavy, and its ninety-horsepower OX-5 was a very low compression engine. He used all the take-off runway there was and dropped almost out of sight down the sixty-percent mountain grade. It was a scary operation, and I didn't blame him for talking about it for months afterwards.

My takeoff was easy because of the higher performing Standard with its 150-horsepower Hisso engine. We both landed on Weeks Field without any trouble.

It isn't my intent to write of my experiences, but instead to give some idea of the progress made in aviation since the early days in the North.

No Winter Flying

I had to discontinue flying in the fall of 1924, for the open cockpits of the Hisso Standards, plus the water-cooled engines, were not suited for the deep cold of Interior Alaska. A decision was made by Rodebaugh to try to get for the company a cabin plane with an air-cooled motor for wintertime use.

Because I was going Outside for the winter to visit my folks in Minnesota, it worked out for me to make a tour of the states to see what kinds of airplanes were being built in the United States. In my travels I learned that about all that were being manufactured were a small number of open cockpit planes with the OX-5 and Hisso motors. One exception was the Huff Deland company which built planes with an open front cockpit for two passengers and a pilot cockpit in the rear. This plane used an early model Wright air-cooled engine of about 200-horsepower. It was unsuitable for Interior Alaska winters—we (Rodebaugh et al) had decided not to settle for anything but a cabin plane, but, as I learned, no cabin planes were then being built in the United States.

The Fokker Cabin Plane

Both the Wright and the Curtiss companies did their best to locate for me a company in the United States that built a cabin plane, but their efforts were unsuccessful.[1]

We finally had to settle on a Dutch-built Fokker F-111, a six-place (pilot and five passengers) monoplane, a model which KLM and early German airlines had flown for scheduled airline service in Europe. It had a 235-horsepower, German, six-cylinder engine. The cabin was plush, with curtains and all the trimmings. It was built in 1921, and it was already the spring of 1925.

The Atlantic Aircraft Company, a dealer for Fokker, had three of these ships available. For $9,000 we bought one of them that

1 The United States lagged behind Europe in the production of cabin passenger airplanes. From 1920 to about 1925, Anthony Fokker, a Dutch builder, produced several models of large cabin planes. The German Dornier and Junker factories also produced multiple-passenger cabin planes; so did the French companies of Caudron and Bleriot; in England Handley Page and De Havilland cabin passenger planes were also being produced.

Scheduled passenger flights by KLM and other early companies with these various makes were common in Europe well before such flights were available in the U.S.

had been used. I arranged for it to be shipped to Fairbanks via the Panama Canal.

This ship proved conclusively that a cabin airplane was the type to use in Alaska, even though we could not use it through the winter of 1925-26 because it had no brakes except for a tail skid, which helped to stop it. It had a rather streamlined monoplane wing, and took a minimum of 1,000 feet to stop after the three points [two wheels and tail skid] were firmly on the ground.

During the summers of 1925 and 1926 we (meaning Noel; no other pilot would fly the Fokker until Joe Crosson arrived, and he apparently flew it as a challenge—not for commercial flights). We [Noel] had some close shaves with the Fokker on sand bars and fields 1,000 feet or under. At the time, our flying out of Fairbanks was the only cross-country flying in the Territory. There was one earlier airline at Ketchikan in southeastern Alaska where for a time Roy Jones commercially flew a two-place Navy training flying boat.[2]

First Flight to Nome

We were successful with the Fokker F-111, and with it made the first commercial flight ever to Nome from Fairbanks, carrying four passengers and 500 pounds of baggage, a 1,200-pound load. We flew non-stop back to Fairbanks in six hours fifty-five minutes.

[Author] *On June 7, 1926, mining engineer Norman C. Stines, with two women, Midge Downer, and a Mrs. Dayo, climbed aboard the Fokker at Fairbanks with Noel and Noel's brother Ralph (who flew as a mechanic) for the 579-mile flight to Nome. The charter price was $1500.*

Noel ran into bad weather along the Yukon River, reversed course and, low on fuel, landed on a steep roughed-out baseball

2 Roy F. Jones, a World War I military pilot, was the first to bring to Alaska an airplane with the intention of starting a commercial airline. His airplane, the *Northbird*, was a four-place open-cockpit Curtiss MF Seagull flying boat powered with a 180-horsepower Hisso engine. In July, 1922, Jones flew this plane from Seattle to Ketchikan, and during the summers of 1922 and 1923, based in Ketchikan, he hauled passengers on joy rides, and mine and cannery owners, fishermen, and others on business trips. The *Northbird* was wrecked beyond repair at Heckman Lake, north of Ketchikan in the fall of 1923. Jones and his passenger, George King, surfaced unharmed. During WWII, Jones flew for the U.S. Army Air Corps from Fairbanks' Ladd Field. He died in 1974.

Noel Wien with the 1921 Fokker F. III which he flew on the first commercial flight from Fairbanks to Nome. This ship was built in 1921 in Amsterdam, Holland. It had no brakes, and carried five passengers.

field at the Yukon River village of Ruby. In the brakeless rollout, the airplane hit a soft spot and turned upside down, splintering the wooden propeller, damaging the rudder and a wing tip. Miraculously, there was no other damage, and no one was hurt.

With help from villagers, the airplane was righted and Ralph Wien repaired the damaged rudder and wing tip. A new prop was hurriedly boated to Ruby from Fairbanks, and the Fokker was ready to fly again. In the meantime, to reach Nome, the passengers embarked on boats.

Noel completed the flight to Nome with the Fokker. It was the first-ever commercial flight between Fairbanks and Nome. The passengers who had fled the upside-down airplane at Ruby arrived at Nome a day after Noel landed the Fokker there.

He flew out of Fairbanks until mid-November, 1926, putting about 100 hours in the air on the two Standard biplanes, and 157 hours on the huge Fokker. He thought A.A. Bennett, the newly-hired company pilot/mechanic, to be so incompetent that he felt endangered whenever he flew planes maintained by him. Bennett apparently convinced owners of the company (Rodebaugh now had partners who had invested in the company) that Noel didn't fly enough. As a result, Noel and his brother Ralph, who worked for them as a mechanic, resigned from the Fairbanks Airplane Corporation. Noel never spoke publicly of Bennett until decades had passed, when the man was long-gone from Alaska.[3]

With Noel gone, Bennett, remembering mechanically-talented 23-year-old Joe Crosson from his time in San Diego, and needing a pilot/mechanic offered him a job. Joe, who had barnstormed in California with his own Jenny, arrived in Fairbanks in March, 1926 and immediately went to work as a mechanic, and soon, as a pilot. (For more on Crosson see Chapter 15).

Crosson, fresh from California, had little idea of Alaska. An advertisement of the time in the Fairbanks Daily News-Miner *might have given him a clue. It read, "Dog team for hire, to go to*

3. In May, 1928, Noel Wien and Russell Merrill with two planes on charter to Barrow ran into difficulties. The two pilots and three other men were missing, and an aerial search was needed. Bennett seemed pleased, even gleeful, that two of his competitors were missing. He demanded $5,000 to fly a search, and then found multiple excuses to postpone the needed flight. Merrill and two passengers nearly died before being found by locals. In the end, all survived, but Bennett lost the respect of many Alaskans.

Later, in Montana, Bennett entered into a business relationship with Bob Johnson, of the famous and highly respected Johnson Flying Service at Missoula. Johnson pulled out after three years, saying, "I couldn't pay Bennett's bills any longer."

Joe Crosson with the Super Swallow biplane NC2375. He arrived in Alaska in 1926 to work as an aviation mechanic, soon became a full time pilot.

Chena Hot Springs or any other place in Alaska. Frank L. Tondro (the Malemute Kid)."

Tondro, a Fairbanks resident, was the genuine Malemute Kid, a relic of the Klondike Gold Rush of '98, made famous by the writings of Jack London.

Pilots accustomed to surveyed and settled country in the states, with roads, railroads, farms and farm fences that helped them to navigate, were sometimes bewildered when they first flew in Alaska where there were no such amenities. Not Joe Crosson. He quickly learned the routes to nearby mining communities and mines. He repeatedly flew to Wiseman, a mining village in the southern foothills of the arctic Brooks Range, a 360-mile round-trip.

In addition to learning the lay of the land by flying over it, Crosson gave flying lessons to Ralph Wien, Ernie Franzen, a mechanic for Fairbanks Airplane Corporation, Cecil Crawford who became a pilot for Arctic Prospecting and Development Corporation, and Andy Hufford, a mechanic for Hubert Wilkins who was then in Fairbanks with his Detroit Arctic Expedition.

With a handful of other pilots, Crosson pioneered aviation in Alaska. He did it the tough way—by flying across a broad swath of the Territory before there were runways, weather forecasts, or radios in airplanes.

Noel Wien traveled to the states by rail and ship, and spent much of the winter of 1925-26 hired but not paid by R. A. Pope, of automobile manufacturing fame, who enthusiastically planned a flight over the North Pole with Noel as pilot. Noel paid his own expenses as he waited for this planned-for but eventually nonexistent event.

Nearly broke, he abandoned the Pope pie-in-the-sky plan, and flew four months with a flying circus in the Midwest, during which time he logged his 1,000th hour in the air. That same year, 1926, he was issued Federation Aeronautique International United States of America pilot's license No. 39, signed by Orville Wright.

Starting Over at Nome
By Noel Wien

[Author] *Following is another "looking back" penned by Noel for the* Wien Arctic Liner *inhouse publication.*

My brother Ralph had been our mechanic (for the Rodebaugh Fairbanks Airplane company) since the spring of 1925, and had helped keep in good flying shape the two Hisso Standards, the OX-5 Jenny that had belonged to Ben Eielson, and the large modern Fokker.

In May, 1927, Ralph and I started a new venture. For $750 we bought from the Fairbanks Airplane Corporation one of the two Hisso Standards that had arrived in Alaska with me in 1924. It was in bad shape. The Hisso engine needed a top overhaul, and the radiator needed replacing. We bought an oversize oval radiator from FAC, and worked hard for about a month on the engine and piecing other parts together. Finally in June, it was ready to fly.

After flight testing for a day or two, I flew it to Nome, arriving June 21. There was no air service at Nome, and after a month or so we found that miners, trappers, and businessmen liked the idea of air service as well as did Fairbanksans. We were a little worried about the engine though, because it was not considered good to have only twenty pounds of oil pressure instead of the normal sixty pounds. Nevertheless, in June, July and August we flew 140 hours, and the engine was still running good.

Noel Wien at Fairbanks with his Hisso Standard in July, 1927, just before flying it to Nome where he flew it until fall cold weather. He carried an extra prop on it at all times.

Miners, traders, salesmen, and missionaries were eager to fly. We provided passenger and freight service to Seward Peninsula mining camps, including Candle, Point Hope, Kotzebue, Deering, and a few isolated mines. Freight had to fit into the tiny front cockpit of the Standard, so most of it was groceries, machine parts, mail, boots, clothing, and the like. We also assisted reindeer herders by searching for strayed animals, saving weeks and dozens of miles of foot travel.

During that time we had some hair-raising flights. We were short on gas in heavy weather, as could be expected from a biplane that cruised at sixty miles per hour. Among our successful flights was one from Nome direct to Anchorage in pouring rain with a low ceiling and no compass, carrying as a passenger George Treacy, a bookkeeper for the Lomen Brothers of Nome. He had gangrene in a leg that needed surgery.

By late August we had netted approximately $4,000 with the Standard. With cold weather near, we knew we had to stop flying soon because of the water-cooled Hisso, and the open cockpits.

The Two Stinson Detroiters

Early that spring (March, 1927) explorer Hubert Wilkins shipped two Stinson Detroiter biplane cabin planes with modern air-cooled Wright Whirlwind 220-horsepower engines to Fairbanks for an attempt at an arctic exploration flight. During a long flight he and Ben Eielson had been forced down on the polar ice, 125 miles north of Barrow in one of these Stinsons [Stinson Detroiter No. 1]. In one of the great feats of Arctic exploration, they left it there to walk over treacherous moving ice to Beachy Point at the mouth of the Colville River.

Wilkins disassembled and stored Stinson Detroiter No. 2 at Fairbanks, and late in the summer offered it for sale for $10,000. He needed the money to purchase a Lockheed Vega, an improved and faster plane for a 1928 expedition.

Ralph and I felt the Stinson Detroiter was the perfect ship for our try at a winter operation. It was the first ship of its kind to be produced in the United States, and the fifth production Stinson built. Its 220-horsepower Wright Whirlwind was the same type of engine Lindbergh had in his Ryan monoplane on his thirty-three-hour flight New York-to-Paris in May, 1927.[4]

Instruments on this ship were modern for their day, with two Pioneer compasses, rate of climb indicator, an airspeed indicator and a turn and bank indicator. Perhaps the most progressive innovations were the inclusion of brakes and a self-starter, which, until then, were unheard of in an airplane.

A total of forty-one of these Stinson Standards were manufactured between 1926 and 1927.

The Detroiter's interior was Spartan. The pilot sat on cushions atop the fuel tank. Passengers sat on their luggage, or whatever was handy. These accommodations, however, beat the interminable Nome/Fairbanks dogsled ride which was the main winter mode of travel before the advent of our scheduled flights.

We purchased Wilkins' Stinson Detroiter in August, 1927, with the profit from the summer's work of the Standard, and a $6,000 loan collected from Nome businessmen in one afternoon.

[Author: The deal included a not-in-writing gentleman's "Alaska style" agreement. Noel and Ralph had only $9,500, and Wilkins wanted $10,000 for the plane. He agreed to accept the $9,500; but if the Wiens succeeded with the plane, they were to pay him the additional $500. The following spring, when Wilkins arrived in Alaska with a new Lockheed Vega, the Wiens handed him $500. The brothers also repaid the $6,000 Nome businessmen's loan.]

This cabin plane with its air-cooled engine was the forerunner of dependable through-the-winter flying in Alaska. We had just the one Detroiter which could operate through the cold months. Beside our flights out of Fairbanks and Nome, we started making

---

4. In 1927, the nine-cylinder radial Wright Whirlwind was one of the most reliable aircraft engines in the world, thanks to seven years of testing for millions of miles of flight by its designer, former racing-car engineer Charles L. Lawrance.

Liquid-cooled engines dominated aviation for years, but they were heavy, bulky, and often cranky. About one-third of forced landings of planes equipped with liquid-cooled engines could be attributed to the engine.

Air-cooled engines eliminated complex liquid cooling systems, which included a pump, pipe-and-hose water connections (as many as ninety on the OX-5), and a radiator to cool the liquid. Elimination of such a system made the Whirlwind at least one fourth lighter, and much less bulky, than a liquid-cooled engine of comparable power.

Early air-cooled engines were inefficiently cooled, and needed a rich (more gasoline) and cool-burning air-fuel mixture to reduce engine temperature. The J-5C Whirlwind's nine cylinders efficiently dissipated heat, allowing use of a lean, and more economical, mixture. This gave an airplane many more miles to the gallon, to use an automobile simile for comparison.

Thus the Whirlwind changed the aviation world with its reliability, efficiency, and lightness. For these reasons, its designer, Charles L. Lawrance, was awarded the 1927 Collier Trophy, at the time America's most respected aviation award.

flights between Fairbanks and Nome. We had bases in both towns, and more and more business developed because of our new plane.

The people of Nome had never had winter service to the Outside (the states) via Fairbanks and Nenana faster than the average time of thirty days by dog team at a cost of upwards of a thousand dollars, plus having to walk, push or run behind a dog sled at least part of the way.

First Alaska Scheduled Flights

Because of the interest shown in air travel between Nome and Fairbanks, we established a schedule. We found we could just about make one round-trip a week, what with the other flying we had in both Nome and Fairbanks. Sometimes it took five or six days to make it one-way due to bad weather, but we managed to average about one flight a week throughout the winter of 1927-28.

One problem we often encountered in Nome was deciding how many passengers could go on a particular flight, and who should wait a week or ten days for the next flight. Sometimes the decision was made by the flip of a coin.

We had a stiff competitor in Fairbanks, [now Bennett-Rodebaugh] but they didn't like the blizzards and the uncertain weather common along the Bering Sea coast and the Seward Peninsula; beside, they had all they could do in the Fairbanks area.

Our weekly flights to Nome from Fairbanks and return were the first scheduled airplane flights between two points in Alaska.

There was a great difference in the weather conditions along the 579-mile route. Fairbanks is in the central interior of Alaska, where calm air is dominant. Nome is on the rough weather, windy coast. The snow at Fairbanks melted two to three weeks earlier than it did in Nome, so take-off from Fairbanks in the spring had to be made from a dry runway. Finding a wheel-landing place in snowy Nome became a real problem during those several weeks.

With the great interest in flying between Fairbanks and Nome, the people in both towns prevailed on the Alaska legislature to appropriate money for three special mail flights, Fairbanks to Nome. Dog team mail could not travel at breakup time, and because of sea ice it was six weeks after breakup before sea-going ships could reach Nome.

We made the first of these three flights during spring breakup, 1928, and we made special air mail flights during spring and fall for the next couple of years. These emergency contracts were the first commercially operated air mail contracts in Alaska, and we made all of them on schedule. Next, came more permanent mail contracts, called Star Route Air Mail Contracts.

Lost Detroiter at Lake Minchumina

For two years, 1927–28, 1928–29, using the one Stinson Detroiter and the same engine, we were able to average a schedule of weekly flights between Nome and Fairbanks, plus many charter flights from both bases. This remarkable performance of one engine might be explained in part by our operating procedure at that time.

We changed oil after every five or ten hours of flying time, thus being able to get just under 1,200 hours on an engine without a major overhaul.[5] An experience I had on Lake Minchumina, near Mount McKinley [Denali], in December, 1927, illustrates the rugged dependability of the old Detroiter.

I was returning from a flight to McGrath on the Kuskokwim River just before Christmas that year. It was important that I get back to Fairbanks in order to make the weekly flight to Nome. Meeting our schedule meant much to us; but to the people of Nome this particular flight was of vital importance, since we would be carrying their Christmas mail, first-class packages and fresh produce. Nome had never had any air service before we started flying from there, but the residents had quickly learned to depend on this modern means of transportation.

I had flown about half the distance to Fairbanks when darkness began to close in; there are, of course, only a few hours of daylight in Alaska's Interior in December. I decided to land on nine-mile-long Lake Minchumina and spend the night at Kamasgaard's Roadhouse. The plane was on skis, and I landed with no trouble and taxied to shore. There I parked the plane in the deep snow of the frozen lake and in the shelter of the trees at the lake's edge.

After a fine meal of roast moose, Kamasgaard and I sat near

5. Even into the 1930s it was common for aircraft engines to need overhauling after 300 or 400 hours of running time.

the fire telling tall tales when I noticed the wind had come up. It was whistling through the trees outside. I didn't worry about the plane immediately because I was sure it was safe in the lee of the big timber, and pretty well settled in the foot and a half of snow. Later, as the wind became stronger, I decided to go out and tie the plane to one of the trees.

It was about eight in the evening when I got to the lake, which was about 300 feet from the roadhouse. I found the snow all blown from the lake's surface, and to my horror, the Stinson was gone. It had been blown completely from sight across the lake.

I started across the ice in the dark to look for it, but the forty-mile-an-hour wind on the glare ice made walking next to impossible. I was forced to give up, and had to crawl on my hands and knees most of the way back to the roadhouse.

Kamasgaard tried to console me by telling me this was the first northeaster to hit the lake that winter, and that they blow up without warning. But I was pretty much disgusted with myself for not having taken the proper precautions with the plane. My only consolation was that I had received another valuable lesson in North-country flying. Neither my host nor I could think of anything we could do that night to retrieve the plane, so we turned in, hoping the storm would abate by morning.

When it became light enough to see next morning, between blasts of blowing snow I was able to see the plane, still upright, about a mile out on the lake. I bundled up in parka and mukluks and headed across the lake. By sliding across the bare ice, and breaking my speed on patches of hard snow, I was able to reach the plane. She was anchored up to the lower wing in the hard-packed, drifted snow. The only apparent damage was a slight buckling on the tips of the lower wing ailerons.

It was obvious the plane would have to be shoveled out, and this could not be done until the wind died down. So, once more, I fought my way through the howling, bitter-cold wind, to the roadhouse.

Later in the day, as I sat at a window watching helplessly, the wind changed direction slightly, and increased in velocity. It undermined snow from the plane and sent that little biplane scurrying off out of sight.

The wind continued to blow at near-hurricane speed throughout that night, and the next day, which was Christmas. What a

The runaway Stinson Detroiter in December, 1927, on Lake Minchumina as it appeared when found nine miles from the Kammisgaard Roadhouse where a storm had blown it. Despite the drooping elevators and other damage, Noel Wien managed to make repairs enough in the field so he could fly it to Fairbanks for proper repair.

way to spend my first Christmas in Alaska, I thought, wondering if we still had an airplane, and how we were going to pay the considerable balance still owed on its purchase price.

It was two more days before that northeaster finally weakened and died, and we were able to venture onto the ice in search of the plane. We found it in the brush on the other side of the lake, deep in snow again, but still intact, and still on its skis.

Repairs in the Bush

It took Kamasgaard and me all day to dig it out, and when we had finished I made a closer inspection for damages. The ailerons on the lower wing were bent a little more, one ski was turned up slightly, and the elevators on the tail were pretty badly crumpled. However, the wings on that sturdy old Stinson had stood up under the terrific buffeting it had taken, and the propeller had suffered no damage. By bracing up the metal ribs I was able to straighten the elevators and the ailerons. The main control rod to the elevators had parted, and I managed a temporary repair. Before I turned in that night I had her ready to fly. And tied down.

Kamasgaard had developed a bad cold while helping me, so he decided to fly to Fairbanks with me. The day was calm and clear. The engine was turning over smoothly when we started our takeoff run. We kept hitting bad drifts, but I figured with the light load I could bounce the plane into the air quickly. But before we got off, the worst happened; one of the skis snapped off just ahead of the pedestal, dug into a hard drift, and nosed us up. Both blades of the prop struck the ice and bent to almost a three-quarter turn before we settled on our ski-and-a-half.

Talk about being discouraged! I was about ready to give up. It was now well after Christmas, and I figured that at this rate I would be lucky to get to Nome before New Year's day.

We spent all that day and the next in getting the plane in shape to fly. Ordinarily, I would have waited for search parties to find me and bring new parts. But it would still be another three or four days before I could expect search parties from Anchorage or Fairbanks.

Kamasgaard and I set about straightening the prop blades by heating them with a blow torch and prying with a monkey wrench. It was a slow and tedious job because the steel blades were tough, and there was the danger of cracking them. With the slightest crack, it would have been extremely dangerous to fly. We spliced the broken ski with a sturdy board, and on the sixth day after I landed at Lake Minchumnina, we were on our way to Fairbanks.

The repaired ski held up for the take-off, and the prop ran fairly smooth, although the engine gained 200 rpm. It seems we had taken some pitch out in the straightening process. This cut our cruising speed considerably, but we finally arrived in Fairbanks, although I was exhausted from fighting the still-bent control surfaces.

It's a wonder that my hair didn't turn gray on that trip. I think I aged ten years. I had made temporary repairs to the main control rod to the elevators by clamping the broken ends together with two pieces of metal secured by an ordinary stove bolt. So long as I maintained forward pressure on the controls I was able to hold the plane's nose down and the rod together. If at any time I had found it necessary to exert strong back pressure, the rod probably would have separated and we'd have plunged to earth. Is it any wonder I was exhausted?

Within two days we had a new ski, and a new propeller installed (it turned out there was a crack in one of the blades of the old prop). We made permanent repairs to the damaged controls, and were on our way to Nome. We arrived there after New Year's day, about two weeks behind schedule, but the rousing welcome we received when we landed more than made up for the anxiety and hardships involved in getting there.

[Author] *The Stinson Detroiter #2 C5262 with which Noel and Ralph Wien flew between Fairbanks and Nome through the winters of 1927-28 and 1928-29, was a key airplane in Alaska's aviation history. That ancient-appearing cabin biplane made possible for the first time year-round commercial aviation for mainland Alaska. During those two winters, the two Wien brothers learned how to keep an airplane flying in deep cold, and many techniques of winter flying—information shared with others in Alaska's infant aviation industry. This experience proved invaluable during the brutal winter of 1929-30 when Alaskan and Canadian pilots flew across the Bering Sea to search the desolate shores of Siberia for the lost plane of Ben Eielson.*

On December 22, 1929, at Fairbanks, while Alaskan Airways mechanic Ed Moore was cleaning the engine of the Stinson Detroiter C5262 with gasoline, a spark set the gasoline and the plane afire. The fire spread. Three other planes in the hangar were yanked to safety, but the Stinson Detroiter #2 and the hangar were both lost to the flames.

4 Alaska's First Bush Pilot

Wrong Font Thompson

Fairbanks, even in the 1920s called a "mining camp" by old-timers, became the center for early aviation in Alaska, partly by chance, partly because of the tremendous advantages of flying, and, helped along by the constant barrage of support of "the aviation" as he called it, from W. F. Thompson, editor of the *Fairbanks Daily News-Miner.* ($2/a month, delivered by carrier).

Nicknamed "Wrong Font," for his initials (he preferred "Wandering Foot"), he loved to write. And he loved to promote. Aviation became a favorite.

Thompson was likable, attractive to both men and women. He liked to drink, and he never worried about money. "He didn't have the faintest idea of a dollar's worth," according to a long-time friend. He was dapper, with a Vandyke beard, and was always immaculately dressed, head back and shoulders squared. He limped and carried a cane because of a poorly healed broken leg from a train accident. He had a world of friends, and few enemies.

Before "the aviation," he promoted Fairbanks as "The Golden Heart of Alaska," even though times were tough in that gold mining town during the early 1920s. When school teacher and former Army pilot (in the reserves, however) Carl Ben Eielson, arrived at Fairbanks in 1922, he often visited Thompson and reporters at the *News-Miner*, which then had both the editorial and printing equipment in a single room.

W.F. "Wrong Font" Thompson, editor of the Fairbanks Daily News-Miner *during the 1920s and early 1930s. He promoted "the aviation" in his newspaper, and even invested in it with Ben Eielson.* News-Miner

Under Thompson's guidance, The *Fairbanks Daily News-Miner* printed everything possible about local airplanes, listing departures and arrivals, names of passengers on bush flights, and the locations of planes down in the bush awaiting repair. The purchase of a new airplane was almost headline news. Thompson glamorized pilots, who, in his eyes, were all heroes. Ben Eielson had no difficulty in talking Wrong Font into investing in the Farthest-North Airplane Company and buying a Jenny, along with Thompson's friend, Dick Wood.

Wrong Font died at age 63, January 4, 1926, at Fairbanks. Among his legacies was the supremacy of Fairbanks as the aviation center of Alaska which lasted for years, in part as a result of his skilled and persistent promotion of, "the aviation."

Another of his legacies was his establishment of the position of Aviation Editor on the *Daily Fairbanks News-Miner* staff. In 1929 and 1930 this position was held by Don Adler. Adler became a student pilot at Service Airlines Flying School in April, 1930.

Following are a few bits from the *News-Miner*'s columns, highlighting and promoting aviation:

> Monday, August 19, 1924. Wien Makes Trip to Eagle in Day. Lands River Bar. Linking Eagle with Fairbanks by air route for the first time, Pilot Noel Wien yesterday made a round flight in remarkable time and with complete success. Leaving the local field at 11:30 a.m. with Norman Wimmler, placer mining engineer of the U.S. Bureau of Mines, as a passenger, he winged his course toward Eagle, arriving at his destination and landing on a sandbar on the opposite side of the river to the town in 3 hours and 25 minutes. Returning, he took a straight course for home, landing

on the local field at 8 p.m. The return flight was made in 2 hours and 45 minutes."

Fairbanks, May 9, 1925. Noel Wien landed afoot in Nenana last evening after walking for three days and nights from where he had landed on a bar, out of gas and out of oil. He briefly stated the facts as above and then announced that he would talk no more until he "had a bath." He's a Cheechako, [newcomer] and couldn't wait until Saturday night.

February 20, 1932: Alaskan airmen again dared death in the Northland when Nieminen and Cope flew to Cook Inlet Sunday and brought Fred "Mulligan" Gotherberg, a trapper who is ill from exposure, to the Anchorage hospital. Gotherberg, believing he was about to die, wrote a letter explaining cause of death. The letter, brought by an Indian, alerted the flyers of the trapper's condition. Obtaining a meager description of his cabin, the flyers searched the Rainy Pass district until they found the cabin. They landed on a snow-covered flat and brought the trapper out to the hospital.

July 1, 1933. Pilot Ed Young of the Pacific Alaska Airways hopped off this afternoon with mail and express for Livengood. Pilot S. E. Robbins arrived in the company's pontoon equipped plane today from Nome with Sam S. Kendrick of the reindeer service as a passenger. Pilot Harry Blunt dropped into the city in another of the company's pontoon equipped planes after a 4,000-mile jaunt around Alaska with Joseph J. Meherin and Lyle Herbert, merchandise, and Chas. Goldstein, fur merchant of Juneau.

The First Alaska Bush Pilot

[Author] *Noel Wien could legitimately be called Alaska's first bush pilot, for he was the first to provide consistent, year-round service for bush residents over a vast region.*

Noel flew for Jimmy Rodebaugh's Fairbanks Airplane Company during the summers of 1924, 1925, and 1926, during which time he established numerous firsts. For several years, planes he flew were often the first to land at many villages and mines. Commonly, residents where he landed had never seen an airplane.

His first flight north of the Arctic Circle (to Wiseman, May

5, 1925) was memorable for its aftermath. On his return flight to Fairbanks, high winds pushed his slow plane far to the south, where he ran out of gas and landed safely on a bar of the Toklat River. He was forty miles from Nenana, and about seventy miles from Fairbanks.

His four-day walk to Nenana from the downed plane, during break-up, became a bush pilot legend. He had a Boy Scout axe, a pocket knife, a Luger pistol, and no food. He arrived at Nenana half-starved and exhausted. He had lost twenty pounds. [And felt he needed a bath.]

Learning How to Fly in Winter

While flying the Stinson Detroiter biplane between Nome and Fairbanks—with charter work at both towns—Noel and Ralph Wien developed procedures for winter flying that soon became widespread in Alaska's fledgling aviation industry. In deep cold, upon an airplane's landing, if it was to remain for a time, oil was drained immediately from the engine and heated before replacing it. In winter, a plumber's pot[1] went everywhere with an airplane. With it was a tent-like canvas engine cover. To heat the engine, the canvas cover was draped over the engine, and the pot was fired up inside it. The cover concentrated the heat on the engine, and was also used to heat the drained oil. This required at least an hour, commonly more, depending on temperature, wind, size of engine, and other factors.

A gasoline-burning plumber's pot used to heat aircraft engines during winter. It sits atop a wood-burning stove. A small tent can be folded inside the stove. If a plane is forced down in the wilderness, the tent and stove could be a life saver. Author

1. They burned gasoline under pressure. Plumbers used them to melt lead used with cast iron pipes.

Airplane cowlings are designed to keep the engine cool while in flight. In deep cold, as -20 F. or -30 F. or colder, engines need less cooling, and Noel and Ralph learned how to modify with baffles and other techniques the flow of air to maintain a warm engine in such temperatures.

The Wien brothers designed new, efficient skis for an airplane. Upon landing a plane, the skis were run up on a pole or board to keep them from freezing down overnight. Wings were sawed with a rope to remove overnight frost, which, if left, spoiled lift, and could keep a plane from flying. Wing covers were soon developed—they could be whisked off the wings quickly, and were easier on fabric covered wings than sawing with a rope.

Airplane windshields were covered overnight to prevent frost from forming. Cold weather lubricants were used; some lubricants hardened in deep cold, and controls, ailerons, flaps, and other working parts didn't work properly with them. A common modification was made to oil tanks by putting a valve and a drain in the bottom, not the side, so they drained quickly and completely.

Lagging was applied in critical areas of the engine. This was a method of insulating by winding asbestos cord around engine parts and coating it with waterglass (sodium silica gel) which sealed it from becoming oil-soaked. (Asbestos is no longer used; various modern materials are now available.) These and many other cold weather techniques kept the Wien planes aloft through long and deep cold winters.

They learned, too, to carry sufficient emergency gear so pilot and passengers could survive an emergency landing in the wilderness.

The Fox Film Expedition

In the spring of 1928, the Wien brothers, Noel and Ralph, bought, for $3,500, from the Bennett-Rodebaugh company the two-cockpit Waco biplane C2775, powered with an OXX-6 (twin magneto) 100-hp engine. Business was good, and they planned to use it for short flights, and for abbreviated landing fields. Noel started to teach Ralph how to fly in it.

While this was developing, in April, Wien Alaska won a contract for $6,000 to fly from Fairbanks to Barrow five men and 2,800 pounds of movie-making equipment for a Fox film expedition.

The Waco would have been useless for this job. Noel hired Russell Merrill of Anchorage Air Transport and his Travel Air biplane to participate.

All went well until the two planes—the Wien Stinson Detroiter, and Merrill's Travel Air—bound for Barrow and beyond the Brooks Range, ran into fog on the North Slope. Both planes were on wheels, and they were forced to land on a tiny frozen, snow-covered, lake.

The Detroiter, with wide tires, handled the snow well. The Travel Air, with narrow tires, bogged down. Next day, with clear weather, Noel took off with the Detroiter and one passenger and flew 115 miles to Barrow, planning to return with shovels and other equipment to get the Travel Air into the air.

When Noel tried to return to the tiny lake, which was one among uncounted thousands of lakes on the vast, flat, North Slope, he couldn't find it. For days he flew frantically, fighting fog and snow, searching.

Merrill and his two passengers remained at the Travel Air for a week, at which time the passengers left to walk for help. Merrill waited a few more days and followed. All three were eventually found, but they were barely alive, starved, and nearly frozen. Merrill especially suffered, and was near death. He was ill for weeks afterward.

At Fairbanks there was no word from either plane, and attempts were made to get a plane from Bennett-Rodebaugh at Fairbanks to fly a search. A.A. Bennett demanded $5,000 for one search flight, and even after agreeing to that, he stalled because of one problem or another. He appeared delighted that his competitor (Noel Wien) was missing, and said that he doubted that the missing men were alive.

Finally, Matt Nieminen, from Russell Merrill's Anchorage Air Transport, flew north with a key transmitter (Morse code radio) so he could report back. Noel and Nieminen then flew a search together, and finally located Merrill's airplane, but by then no one was there. Nieminen's radio was unable to reach Fairbanks, or even Wiseman, so for weeks those in Fairbanks had no word the lost planes and the men who had been in them.

It all ended well, but it was a harrowing experience for all involved—including those waiting at Fairbanks.

Ralph Wien's Mail Flight

That spring of 1928 Wien Alaska Airways had a contract to make three mail flights from Fairbanks to Nome. The third and last flight was scheduled for May 23. Noel was missing on the North Slope when that date rolled around. Ralph Wien and the newly purchased Waco were the only resources of the Wien company then available.

Ralph had only eight hours of dual instruction as a pilot, and had flown solo for a scant two hours. He had never made a cross country flight alone, had not figured out any navigation problems, although he had frequently flown with brother Noel on long flights. He had landed an airplane only on the relatively long and smooth Fairbanks Weeks Field.

If Wien Airways failed to make the May 23 mail run, it could have lost the mail contract. Ralph, despite his lack of flying experience, decided to fly the mail to Nome. He would honor the contract.

When the first class mail for the Seward Peninsula arrived in Fairbanks by ship and train, he loaded the 500 pounds into the front cockpit of the Waco, and, despite having a heavy cold verging on pneumonia, climbed the little biplane into the Fairbanks sky and headed north.

The route called for eleven landings, some on rough, short runways. Ralph made every one on schedule, and nonchalantly delivered the mail as if he did it every day. On his return to Fairbanks, when asked if he had any problems on the flight, he replied succinctly, "None."

Later, brother Noel, and another veteran bush pilot, Bob Reeve, called the flight, "One of the greatest flights ever made in Alaska."

Ralph flew commercial flights with the Waco from Fairbanks until Noel's return on June 16.

Wien Headquarters at Fairbanks

For two years Wien Alaska Airways had provided service year around between Nome and Fairbanks. Now Noel and Ralph decided to headquarter at Fairbanks, and in early 1929 they built a big new hangar at Fairbanks' Weeks Field. Their "fleet" of airplanes included the faithful old Stinson Detroiter No. 2 biplane C5262, a grand new Hamilton Metalplane NC10002 (see following), the Waco biplane Model 9 C2775 which they bought in early 1928, and a soon-to-be-purchased Stearman biplane C-2B NC5415. The company pilots were Noel and his brother Ralph. In July they hired pilot Frank Dorbandt and stationed him at Nome. Dorbandt had arrived in Anchorage the previous April to fly for Anchorage Air Transport.

In October, 1928, there were eight licensed pilots in the Territory of Alaska, of which six were at Fairbanks. There were seventeen airplanes (not all used commercially) in the Territory, and twelve licensed aircraft mechanics, of which nine were in Fairbanks. "The aviation" in the Territory was still a struggling, but clearly a growing, infant, and Fairbanks was its center.

The Hamilton Metalplane

The Wiens continued to fly the Detroiter cabin biplane through the spring, summer, and fall of 1928. By the end of 1928 Noel had logged more than 1900 hours of time in the air, most of it in Alaska. He had the most air time in Alaska of any pilot, and was among the most experienced Far North aviators in the world.

In late fall, 1928, a smallpox epidemic struck Native villages in northern Alaska. With the Stinson Detroiter cabin biplane, Noel flew Dr. J. A. Sutherland of Fairbanks to check it out and administer vaccine at various villages, including Ruby, Russian Mission, Marshall, Mountain Village, Kaltag, Unalakleet, and Nome. On his return to Fairbanks in early January, waiting for him Noel found the Hamilton Metalplane, ordered months earlier. The Wiens were in good company with that modern airplane; Northwest Airlines had ordered eight of the new Hamiltons for their passenger service.

The December 28, 1928, *News-Miner*'s announcement of the airplane's arrival read, "The task of assembling the giant Hamilton monoplane of Wien Airways which arrived in pieces on the freight train yesterday will take five days or so say mechanics for the company."

The *News-Miner*'s detailed report continued enthusiastically: "The most striking features of the craft are its size and luxury. The fine quality of its fittings and its suggestion of comfort make apt

Wien Alaska Airways purchased the Hamilton Metalplane NC10002 (right) in late 1928 after flying the Stinson Detroiter C5262 (left) between Nome and Fairbanks for two years. Both planes were in Fairbanks in the summer of 1929 for this photo.

the name of 'air yacht.' Nothing that might make for the ease of the passenger has been overlooked.

"In the passengers' compartment, separated from the pilots' cockpit by a wall which aids in deadening the noise of the motor, are six wicker chairs, three on a side. Passage can readily be had through a roomy aisle. The chairs are fitted with deep cushions. The doors opening into the compartment from the outside are built on the same plan as those in enclosed automobiles, with windows that shut or lower at will, and with handy pockets. The doors are finished in mohair.

"In the rear of the passengers' cabin is a lavatory containing a toilet, water tank, soap receptacle, and wash basin. In this compartment there is also room for baggage. Entrance to it is gained through a door, and at the other end of the lavatory is a door which opens on a long passageway which can be used for storing furs or other bulky material.

"The pilots may reach the cockpit either through the passenger cabin or through a window directly over the cockpit. Seats and controls for two pilots are provided.

"The all-metal construction of the monoplane brings to it an inherent safety factor which is being adapted to an increasing extent by airplane manufacturers. The metal is light yet strong, precluding any possibility of splintering in event of a crash.

"'The only non-metal parts of the plane,' Ralph Wien said, 'are the tires and mohair cushions and lining and the mohair is fireproof.'

"There is a heater in the cockpit, and two in the cabin for passengers. In the latter there are two dome lights. Mounted on the wing are big navigating lights and the pilot has dash lights to make his instruments visible at night. All the conventional instruments are carried. The compass is mounted behind the pilot and is read through a rear vision mirror. In the rear of the fliers there is also a control for adjusting the stabilizer.

"The motor, a 425-horsepower Pratt & Whitney Wasp, is the most powerful ever brought here to be used for commercial service. It is certified and has the reputation of being one of the most reliable engines available.

[Noel later replaced the Wasp with a 525-horsepower Pratt & Whitney Hornet.] "The big plane is equipped with a self starter. One hundred and forty gallons of gasoline, sufficient for seven and a half hours of flight, can be carried in the tanks, which are placed in the wing.

"The landed cost was over $26,000, and it will be flown by Noel Wien. The plane will be the largest commercial plane in service in Alaska."

The Hamilton Metalplane[2] NC10002 was the first new airplane purchased by the Wiens, and it was to become one of Alaska's most famous airplanes. The company later acquired one other, and possibly, two Hamiltons.

That summer, all flights of the new Hamilton carried a flight mechanic, and Earl Borland was often that mechanic. Borland, in addition to being a mechanic, was learning to fly. On July 17, on a flight with the Hamilton from Nome to Fairbanks, Noel allowed him to pilot the Hamilton during much of the way.

On September 23, 1929 Borland made his first solo flight and three landings at Fairbanks (not with the Hamilton). His instructor was Ralph Wien.

Today (2009) there are only two known surviving Hamilton

---

2. The Hamilton Metalplane was the first new airplane purchased by the Wiens. It was manufactured at Milwaukee, Wisconsin, by the Hamilton Metalplane Company, which became a division of the Boeing Airplane Company in late 1928. The airplane had an all-metal aluminum alloy framework, covered with a corrugated dural metal skin. Top speed (with the later installed 525 hp Hornet) was 145 mph, service ceiling 16,000 feet, range 600 miles. Height, 9 feet 4 inches; wing span 54 feet 6 inches. Empty weight was 2,450 pounds; payload 1,100 pounds.

The Wien Stearman open-cockpit ship at Walker Lake, wings newly repaired by Earl Borland and Fritz Wien, ready to fly to Fairbanks about May first, 1929. This is the airplane flown by Harold Gillam when he and Joe Crosson found the lost Hamilton NC10002 in Siberia. Today, this Stearman, in airworthy condition, is one of the most prized airplanes in the collection of the Alaska Aviation Heritage Museum in Anchorage.

Metalplanes. One is owned by the Alaska Aviation Heritage Museum in Anchorage. Wien Airlines purchased it in 1937, and it was damaged in 1939 when Sig Wien, Noel's brother, made a hard landing with it at Solomon, Alaska. The company decided it wasn't worth repairing, and it ended in Iowa and then California. The Alaska Aviation Heritage Museum in Anchorage bought it with a $34,050 grant from the Anchorage based Rasmuson Foundation.

It needs much restoration work. The prop, instruments, control wheel, and rudder pedals are missing. The fuselage is on a wall at the museum, with plans to attach one wing. The engine is in a museum hangar. There is not enough money to restore the airplane at present.

The other surviving Hamilton has been totally rebuilt to airworthiness, and is in a Minnesota museum. It too has an Alaskan provenance.

The Walker Lake Stearman

The Stearman NC5415 the Wiens bought in March, 1929, had been purchased from Varney Airlines at Boise, Idaho, by Arctic Prospecting and Development Company of Fairbanks. It was

Ralph Wien and his wife Julia (far left, and far right) with their children Bobbie and Jimmie. Noel and his wife Ada are in center. Behind them is the Hamilton Metalplane NC10002. Circa 1929.

Early planes did not have self starters and had to be started by turning the engine with the propeller, a dangerous practice. Here the man twisting the prop is pulled to safety by the others when the engine of this biplane starts.

Noel Wien and the fifth Stinson built, brought to Alaska by explorer Hubert Wilkins and flown for him by Ben Eielson. It was purchased in 1927 by Noel and Ralph Wien. The Wiens flew it between Nome and Fairbanks for the first scheduled air service in Alaska. It was the first air-cooled engine, closed-cabin airplane to fly commercially in Alaska. It used the nine-cylinder, Wright Whirlwind 220 hp engine. Photographed at Nome in October, 1927.

shipped from Seattle to Anchorage where it was assembled in the shop of Anchorage Air Transport. Cecil Crawford flew it to Fairbanks on July 16, 1928, and from there to fourteen-mile-long Walker Lake, at the head of the Kobuk River, and ninety miles northwest of Bettles.

Crawford briefly flew freight with it to and from the lake. On July 21, 1928, while landing, he struck a bump and bounced into the air, then struck another bump. The landing gear broke and the plane turned onto its back. A wing was damaged and one prop blade broken. It remained where it fell.

In early February, 1929, Noel flew to Walker Lake to inspect the plane with the possibility of purchasing it. It appeared repairable, and the purchase was made for $2,500. On April 16, Ralph Wien, with the Stinson Detroiter and newly hired mechanic Earl Borland, flew to Walker Lake. Borland had been a garage foreman for the Alaska Road Commission. In 1927 he had left the Road Commission to work on airplanes for the Bennett-Rodebaugh Company. Now he worked for the Wien company.

The Stearman needed a new engine mount and a new wing panel. The panel had to be built at Fairbanks. Ralph and Borland took measurements, and Ralph returned to Fairbanks to do the work there. Borland remained to work on the plane while living in a nearby cabin.

Two weeks later, Noel and Ralph, flying the Hamilton, with their brother Fritz, who was newly arrived from Minnesota and now working for the company as a mechanic, arrived at Walker Lake with the new wing panel and other parts. Fritz had with him welding equipment needed to complete the repairs.

The four swiftly completed the repairs. Ralph was to fly the Stearman to Fairbanks, but Noel decided to test hop it first. When the plane lifted off he quickly discovered the aileron controls were reversed. When he moved the control stick to the right as in a right turn, the airplane banked left, and vice versa. This would be like with turning the steering wheel of an auto to the right, only to have the car veer left, and vice versa.

He managed a safe landing.

The aileron cables had been crossed when replacing the wing panel. Long term aviation records show this problem crops up occasionally, and that dead pilots can result. Noel's quick understanding of the problem saved the airplane, and perhaps his life.

The Walker Lake Stearman, commonly called the Wien Stearman, was to make history. It still exists, on display at the Alaska Aviation Heritage Museum in Anchorage, as one of the most treasured of the museum's many airplanes.[3]

(For more on Noel Wien, see chapter 19, "Noel Wien, The Later Years.")

---

3. Wien Alaska Airways flew the Stearman for about five months; it was included in the sale to Alaskan Airways in September, 1929. Harold Gillam flew it on the search for Eielson. Alaskan Airways merged with Pan American Airways under the name Pacific Alaska Airways, for which the Stearman flew for many years. Later, while flying for Cordova Airlines, it crashed in the Dadina River valley near Mount Wrangell, and languished there until the 1960s. The recovered remains were acquired by Les and Janet Kares of Stevensville, Montana, who spent ten years meticulously restoring it from the original fuselage, the Whirlwind J-5 engine, control stick, rudder, brake pedals, cowling and original brass data plate. Wings and other components were built from scratch.

5 Olaf Swenson, Trader

IN THE SPRING OF 1901, 18-year-old Olaf Swenson and his father, Big Nils, joined the gold rush to Nome where, through the summer, working fifteen hour days, each made about thirty dollars a day. Born in Manistee, Michigan, in 1883, Olaf was a powerful six-footer even as a teenager. His six-feet-two-inch 260-pound father had emigrated from Sweden.

The Northeastern Siberian Company Limited, financed by English, French, American, and Russian capital, held a concession for mining and trading in northeast Siberia, from the Anadyr River to North Cape. In early 1902, this company grubstaked fifty Americans, including Olaf and his father, and to have them look for gold, transported them from Nome to St. Lawrence Bay, Siberia. They were provided with lumber and tools to build a boat, which they sailed to Kolyutchin Bay in the Siberian Arctic to prospect.

They failed to find gold that summer and arrived back at St. Lawrence Bay too late to catch the company's south-departing steamer. They wintered there. Gold prospecting spring, summer, and fall of 1903 also failed, and Olaf and his father returned to Seattle, where, until 1905, Olaf worked as cashier and credit man for the American Biscuit Company. He also married, and soon he and his wife had a son. The challenges of the Arctic had captured him and he yearned to return there; selling biscuits bored him.

Return to Siberia

Olaf theorized that gold deposits in Siberia, fifty-one miles across Bering Strait from Alaska, must be similar to those at the Bering Sea coastal town of Nome. He approached the Northeastern Siberian Company with this theory, and at no cost they issued him and a group of about twenty-five other young men he had assembled a concession on about 165,000 square miles in Siberia (an area slightly larger than California) where they could stake claims. If they found gold, they were to get half, plus half the gold-producing ground; the other half of the gold and ground would go to the Company.

These adventurers sailed from Nome aboard the company-owned motorship *Barbara Hernster,* bound for the Anadyr region in northeastern Siberia, with provisions, mining equipment, and dynamite for the prospectors. The company had exclusive trading concessions at Anadyr. Count Podgorsky, a Pole, an executive for Swenson's company, was also aboard with a large supply of trade goods for the Chukchi Eskimos of coastal Siberia.

Olaf, now 22, was accompanied by his wife and their six-month-old son.

While sailing in dense fog a half mile off of the Siberian coast the *Barbara Hernster* hit, and became stranded on a rock. Everyone aboard arrived ashore safely in the ship's small boats. Next morning's tide lifted the *Barbara Hernster* clear of the rock and, by setting sail, the crew managed to beach the ship. The cargo was water damaged, but many items were salvageable. The ship was a total loss.

Olaf Acquires Trade Goods

Olaf proposed to Count Podgorsky that he salvage the trade goods in exchange for some of the items. Podgorsky agreed. Olaf paid men in the party a dollar an hour to salvage the cargo. In exchange, he received sacked flour and bricks of pressed Russian tea and leaf tobacco. Podgorsky retained the canned goods; labels had been soaked off, and one might get sauerkraut when peaches were desired, but the contents were still edible.

The sacks of flour were wet on the outside, dry inside. Olaf dumped the dry flour into a wooden bin he hired local Natives to build, and refilled the dried sacks. He spread the tea and leaf tobacco in the sun to dry and repacked it. Both were salty, but usable, and valuable for trading.

The engineer and two crewmen from the *Barbara Hernster* sailed 190 miles to Nome in a whaling boat and arranged for another company ship to pick up the beached party and their goods and take them to Anadyr, several hundred miles up the Siberian coast.

Anadyr, and Trading with Eskimos

Anadyr was a bleak, remote wilderness with sparse vegetation. The land is continuously wet in summer and dotted with lakes. The region is a true desert—a desert of frozen land. The wealth of the area is derived from furs of the land mammals, and the fish, seals, walrus, seabirds, and polar bears in the Bering Sea. A rare prime silver fox skin, a color variation of the red fox, brought anywhere from $500 to $1,500. A white arctic fox skin, the most abundant fur, was worth about $50.

Summer temperatures range from +36 to +50 F.; winter temperatures from -16 F. to -21 F. with extremes of -49 F. Precipitation averages about ten inches, including 36 inches of snow.

This region was thinly populated with tiny and widely-scattered villages of Chukchi Eskimos, small, dark-skinned people with an oriental cast. There were no roads, only trails. Transportation was by dog team or reindeer during the long, harsh, snowy winters, and by foot and skin-covered kayak during the three or four snowless months.

The prospecting party built a station at Anadyr. While the other members of the party headed inland to prospect, Olaf and his wife remained on the coast, living in a tent, and trading the salvaged flour, tea, and tobacco for the furs of local Eskimos.

The Russo-Japanese war (1904–5) was under way, resulting in a shortage of supplies in Anadir. When Olaf and his party arrived, flour was selling for eleven dollars a sack. Not wanting to profit from the situation, and looking to the future, he sold his flour for a dollar seventy-five a sack, and buyers flocked to him from miles around. The salty tea and tobacco also traded well.

The considerable profit Olaf made in trading his salvaged goods for furs that year financed his start as a long-time trader along the Siberian coast. A year after trading all his salvaged trade goods for furs, he returned with a large and more varied supply of trade goods. About then, the Tsar annulled the concession of the Northeastern Siberian Company. That had little or no effect on Olaf; by then he was nicely established as a trader of American goods with Siberian Eskimos.

The Siberian Eskimos

The Siberian Eskimos Olaf traded with were of two cultures; interior and coastal. The interior Eskimos depended on reindeer they herded for food and clothing. Their houses were pole frames, over which reindeer skins were stretched. Inside, small rooms were formed by skins hung from ropes. Stone seal-oil lamps heated small sleeping rooms. A hole at the top of the house allowed smoke to escape from the ever-present heating/cooking fire.

Coastal Eskimos lived from the sea on seals, fish, walrus and polar bears. They lived in houses about sixteen feet square, built of driftwood. The entrance was commonly a storm shed. A thirty-feet-long tunnel then led to the house proper, where an open heating and cooking fire burned. They too had a hole in the top for smoke.

Interior Eskimos made smoking pipes from alder root, the bowl lined with lead, and the mouthpiece made from an empty rifle cartridge. Before loading the pipe, a pinch of reindeer hair was stuffed into the bowl, with the tobacco on top. It was then lit, and kept burning by use of a tiny bone or a piece of wire used as a poker. The goal of smokers seemed to be a spell of heavy coughing—the more violent the cough, the better. Many of their pipes were huge, weighing up to five pounds. They were carried in a buckskin bag slung from the shoulder.

Coastal Eskimos used smaller pipes, some made of petcocks from gas engines for the bowl, with copper tubing for the stem.

All Natives chewed tobacco, including women and children. A chew was commonly passed from person to person; a house guest expected to be offered a chew, even though the quid was already being chewed by the host. A crying baby was commonly quieted by its mother who would take a chew from her mouth, or from behind her ear, and stuff it into the child's mouth. The crying usually stopped.

The Eskimo houses, both coastal and interior, were warm even in the deep cold of winter, but to Olaf, the odors in them were almost unbearable. Despite the strong smell, the hospitality was generous and universal. Strangers arriving at a village at night were welcome at any house, where sleeping space and food was always available.

There were few elderly Siberian Eskimos. In time Olaf learned that when the aged became entirely dependent upon others, or had a terminal illness, they asked a close male relative to send them to their Spirit World. Strangulation was the usual method to escort these elderly or ill out of life. In more prosperous villages, they could be sent on their way with a gunshot.

One of Olaf's hired Eskimos asked for a rifle and other goods on credit, which he was given. Next day he told Olaf he had killed his wife, who was in the end stages of tuberculosis. He placed her and the rifle on the tundra. Months later he killed a demented woman, and also put her body on the tundra. His mother-in-law still lived with him. She asked him to let her follow her daughter, to which he agreed. He plunged a knife into her heart to grant her wish to end her days.

Olaf knew the man to be kind and not at all evil, yet within a couple of years he had killed three women. Olaf found it difficult to get used to such customs, but eventually he decided they were about as sensible as some of the customs of our so-called civilized societies.

Over the years Olaf owned and used a number of ships to sail to the Siberian coast for trading. In 1913 he purchased the *S.S. Belvedere*[1], a steam-powered former New England-based whaler. That fall, after leaving the Siberian coast on his first trading trip with the *Belvedere,* he sailed to Nome, where the *Belvedere* picked up freight to deliver to Herschel Island for explorer Vilhjalmur Stefansson, as well as freight for the mounted police station at Herschel Island. On the way, the ship became stuck in the ice near

1. Built in Bath, Maine, in 1880, the *Belvedere* was a 3-masted, 440-ton bark. She made her maiden voyage in 1881 as a whaler to the Western Arctic. After that she made 24 more whaling voyages from San Francisco into the Arctic, before being converted to an arctic trader and freighter in 1909. The *Belvedere* was crushed by ice and sank off Siberia in 1919.

Demarcation Point, sixty-five miles from Herschel Island, where she remained for the winter.

Swenson left the ship and made his way overland to Fairbanks and on to Seattle, where he outfitted another ship, the *King & Winge,* with supplies for the *Belvedere,* and the usual trade goods for Siberia. He planned to deliver the supplies to the *Belvedere*, and then to carry on with his annual trading voyage to Siberia with the *King & Winge*.

It was then that he became entwined in the rescue of marooned members of Stefansson's 1913-18 Canadian expedition. It started when, in August, 1913, the fragile 250-ton *Karluk*, the primary ship of Stefansson's expedition, became locked in ice north of Alaska. It was carried toward Siberia. In September, Stefansson and five other members of the expedition left the ship to go ashore to hunt caribou. That was the last those aboard the *Karluk* saw of Stefanson and his companions. Most of the twenty expedition members remaining on the *Karluk* were convinced that Stefansson deliberately abandoned them, knowing the ship was locked in ice and being carried away from the Canadian Arctic where he planned to work.

Ice carried the *Karluk* hundreds of miles west to near Russia's Wrangel Island, where, damaged by ice, she sank in the Chukchi Sea, leaving the expedition members on the ice with what they could salvage while the ship was sinking.

Dissention caused the twenty survivors to split up. Some headed one way, others another. Most of those who were led to Wrangel Island over broken and floating ice by their experienced captain, Bob Bartlett, survived, but, of the twenty, eleven men died.

Bartlett, with an Eskimo companion, made a hazardous 200-mile trip across shifting ice from Wrangel Island to the Siberian coast. Next they traveled, afoot, 700 miles south in Siberia. From there the whaler *Herman,* owned by Captain C. T. Pedersen, took him to Nome where he notified the world of the marooned expedition members on Wrangel Island.

The U.S. Coast Guard ship *Bear* attempted to reach the Russian Island for a rescue, but ice and a low fuel supply aboard thwarted the attempt. Bartlett appealed to Olaf Swenson, who was in the vicinity with the *King & Winge* delivering supplies to the *Belvedere,* which was now free of the ice.

In the spring of 1914, Swenson diverted the stout *King & Winge*[2] from his trading route to Siberia, and worked his way through ice to reach Wrangel Island to rescue the nine surviving expedition members.

Within a few years Olaf learned much about trading with the Chuckchi people. To make his flour, sugar, pilot bread, clothing items and other goods attractive to these Eskimos, Olaf wrapped them in waterproof paper, oilcloth, and burlap, making the packages of uniform size, and each of approximately the same value. The packages easily fit in the common back-pack used by the people, as well as in sleds they used in winter.

Olaf's thriving business with the Chuckchi of Siberia ended with the 1917 Russian revolution, when the trading posts he had established were confiscated.

By then trader Swenson spoke Russian, as well as the language of the Chukchi Eskimos. In 1926, after much effort and time in Russia dealing with bureaucrats, he entered into a five-year joint trading venture with the Soviet government.

The *Nanuk*

Swenson agreed to deliver to Siberian coastal villages specified American merchandise. In exchange, the U.S.S.R. was to deliver to him fur from an area larger than Alaska to be sold on fur markets in New York and London. The Swenson Fur Trading Company received a percentage of the return from the furs. According to Swenson, during the years of this contract the Soviets faithfully lived up to their agreement.

Swenson planned to leave Seattle each year with a ship loaded with trade goods that sailed northwest 4,000 miles to the Siberian coast and the Kolyma River. On his return to Seattle he expected

---

2. The *King & Winge* (named for shiprights Tom King & Al Winge and their shipyard) was one of the most famous ships ever built in Seattle. Launched in 1914, and built originally as a powered halibut schooner, this 110-foot ship had 4x4 1/2-inch oak frames six inches apart. She was sheathed with three-inch-thick planking and covered with a layer of ironbark. Her bow was originally covered with steel plates for ice work. On October 25, 1918, she was nearby when the *Princess Sophia* sank on Vanderbuilt Reef in Southeast Alaska and 398 lives were lost. She was used as a Columbia River bar pilot boat 1924-58. During the late 20th century she worked as a king crab fishing boat and a commercial fishing longliner in Alaska. She sank February 23, 1994, in eighteen-foot seas, 22 miles west of St. Paul Island, Alaska. Her last owner was Richard Maher, of Homer, Alaska, a friend of the author. —JR

to have a load of valuable fur brought from inland and collected from stations along hundreds of miles of the Siberian Coast. The fur went by rail to New York.

The ship he initially used for this was the *Nanuk,* a former Pacific Coast lumber schooner, which he bought from Northern Whaling and Trading Company in Seattle. Built by Hans Bendixsen in 1892 in Eureka, California, this three-masted schooner (a schooner is a ship usually of two masts, with the smaller mast and sail forward) of 120-feet length, was first named the *Ottilie Fjord.* In 1923 the Northern Whaling and Trading Company purchased her and re-named her *Nanuk*, Alaska Eskimo for polar bear. Swenson retained that name.

For protection from arctic ice, the hull of the *Nanuk* was sheathed with two-inch-thick Australian ironbark, one of the hardest woods known. She had a small cabin aft, as well as the wheel, and was powered by a 200-horsepower, six-cylinder Atlas Imperial diesel.

In both 1926 and 1927 the Nanuk sailed the round trip to Siberia without difficulty, working her way through scattered ice, and spending about three months each trip. Fur was brought from hundreds of miles inland for Swenson to pick up on the coast. The two successful trips added to Olaf Swenson's capital, for fur prices were high, and the harvest was huge. It seemed Olaf had found a way to riches. But, as he was to learn, it wasn't always simple, or easy.

The *Elisif*

In early 1928 Swenson signed another contract with the head of the Soviet Fur Trust, which called for a second ship suitable for the Arctic. In Norway he bought the *Elisif,* another schooner. Her English-speaking Norwegian crew, captained by Evan Larsen, of Brevik, Norway, sailed her across the Atlantic, through the Panama Canal, then to Seattle. She was ready for his use for the 1928 season.

That spring the *Nanuk* and the *Elisif* left Seattle together, headed for Siberia. Before leaving Seattle, a radiotelegraph (Morse code—no voice) was installed on the *Elisif.* The two ships encountered severe ice conditions in the Chukchi Sea north of Bering Strait, and the propeller of the *Nanuk* was damaged by ice. She

In March, 1929, the frozen-in-ice trading schooner Elisif *at North Cape, Siberia, held a cargo of valuable fur. "From a distance about all that showed were the masts. Snow had drifted over the decks and extended leeward three or four hundred feet," Wien wrote. With the threat of falling fur prices, the owners contracted to have Wien fly some of the fur to Alaska for shipment by rail and ship to the New York fur market.*

returned under sail to Seattle for repair. Before leaving the North, most of her cargo was stored at Teller, Alaska.

Locked in Ice

The *Elisif* continued north along the Siberian coast, trading, and gathering furs at designated collection points. On August 22, 1928, headed south, bound for Seattle, she was ice-locked in the open sea about eleven miles from North Cape[3], Siberia, where one of the Soviet stations held furs ready to be picked up. Natives arrived with a skin boat (which could be dragged across the ice between areas of open water) with a letter from Kavelin, the Soviet

3. Since renamed Mys Smidta (Cape Schmidt) after famed Soviet explorer Dr. Otto Schmidt.

representative at the station, listing nearly 8,000 furs of white and red fox, polar and brown bear, hair seal, and reindeer fawn, as well as 1,610 pounds of walrus ivory. He wanted Olaf to travel to the station to inspect this shipment.

A week or so later the crew of the *Elisif* managed to break her free from the ice, and move about eight miles closer to North Cape, where she was again locked in ice within three miles of shore. This time the ship was stuck for the winter.

Olaf Returns to Seattle

Swenson, who was aboard the *Elisif*, needed to return to Seattle. The *Nanuk,* under repair in Seattle, needed to be reorganized for the following season. There were two travel possibilities; he could arrange to charter an airplane to fly him the 450 miles from North Cape to Alaska; or he could travel by land to Europe and take passage on an ocean liner to New York.

Through the *Elisif's* radio he communicated with Fairbanks-based Wien Alaska Airlines, learning the cost of chartering an airplane from Alaska was astronomical. Further, no one had ever flown an airplane between North America and Asia. Because he believed a flight across Bering Strait was extremely hazardous due to high winds and the commonality of blizzards, he decided against chartering an airplane.

Olaf had fond memories of the fabulous trip he had made sixteen years earlier, in 1913, when his *S.S. Belvedere* became frozen in the ice of the Canadian Arctic. In twenty-six days, he and Captain C. T. Pederson, captain of the whaler *Herman*, (and father of now deceased Ted Pederson, a long-time resident of Homer, Alaska—author), mushed two dog teams 700 miles to Fairbanks from the *Belvedere.*

The two men crossed the Arctic Brooks Range at Chandler Pass, paused at Fort Yukon, and Circle, and hurried on to Fairbanks. They traveled fifty-three miles on their last day, starting at 5 a. m., and arriving at Fairbanks at 11 p. m., walking and running the entire way beside their loaded sleds, which carried more than 300 white fox skins.

Olaf had enjoyed this trip because of its adventure and the

beauty of the snow and ice. Frost on the trees created a fairy-land; when he bumped such a tree, crystals showered in a sparkling fall. After the life he had known in the Arctic, near the sea and its constant wind, the typical still, dry air of Interior Alaska made travel most enjoyable. Now, sixteen years later, another winter dog team trip appealed to him. This time it would be across Siberia.

So, for the second time in sixteen years, Swenson left an ice-trapped trading ship (the *Elisif*) to travel to Seattle to get a ship (the *Nanuk*) ready for the coming trading season, just as he had left the *Belvedere* in 1913, to travel to Seattle to prepare the *King and Winge* for the 1914 trading season.

On October 19, 1928, with Kavelin, the North Cape Soviet official, Swenson left the trapped *Elisif*, and undertook one of the most remarkable modern-time arctic journeys on record. Swenson estimated that he and Kavelin traveled 4,100 miles zig-zagging across the vast, thinly inhabited land of tundra, frozen rivers, and rugged mountains to arrive at Irkutsk, a town and station on the Trans-Siberian Railway. Straight-line distance North Cape to Irkutsk is about 2,100 miles.

Before leaving, he had a Native woman sew him a fur parka and sealskin pants that were lined with Hudson's Bay wool blanket material, as well as mittens and socks.

For overnight housing and shelter from blizzards, he and Kavelin depended on the hospitality of coastal and interior Siberian Natives. They started with a fine dog team, and as those animals tired, they were exchanged for fresh village dogs. Later, they traveled by reindeer-pulled sleighs. Nearer Irkutsk, they rode in horse-drawn sleighs and carts.

They fought blizzards, endured deep cold, and puzzled over thousands of miles of confusing trails. It was an epic accomplishment.

One of the first stops on this long journey was at the port of Nizhni Kolymsk, the most important trade distribution point on the Kolyma River. The *Elisif* had been unable to reach this port, so there was a shortage of foodstuffs and supplies. Furs had been collecting there for a year, awaiting the *Elisif*. The collection included 119,736 skins of squirrel, ermine, red, cross, and silver fox, white and blue fox, otter, wolverine, wolf, reindeer fawn, plus walrus ivory and petrified mammoth ivory.

Living conditions of the Natives with whom they stayed were

basic. With two Natives, temporary traveling companions, they were held up three days by a blinding blizzard, during which they lived in an unusually large native house of about twenty-five feet in diameter. It had a living room, dining room, kitchen, sleeping room and toilet.

Though by local standards it was a grand house, Olaf and one of his companions, because of the almost unbearable aroma in the main section, preferred to sleep in the outer, unheated. part. They were warm in wolf-skin sleeping bags, even at -30 F.

The woman who lived in the house was softening reindeer leg fur to make mittens. She chewed the skin, dampening it with water. She then scraped it with a curved piece of wood which held an iron washer.

Her seventeen-year-old daughter, naked except for abbreviated tights, (the Native people had no idea of European physical modesty) hung ice-filled kettles over seal oil lamps. When the kettles boiled, tea was made.

This young woman used a dirty cloth to wipe cups before pouring tea into them; if there was a spot, she spit on it, and used the cloth to wipe.

Dishwashing was simple; the girl licked them clean. Tied sled dogs licked the meat platter clean.

While Olaf and his companions were guests in this house they were served raw frozen meat twice a day. It was sliced thin, and dipped in seal oil. Before going to sleep at night, they had a small helping of boiled meat. The Natives ate with fingers and knives; holding meat between their teeth, and the other end with fingers, they passed a knife between lips and fingers.

Fish fueled much of Olaf's great journey. It fed the sled dogs, as well as the travelers. He was surprised to learn that horses and cows in Kamchatka love fish; he once watched a horse steal a live fish from a man who had just caught it. The horse held the fish's head in his teeth and ran off, with the tail of the fish still flapping.

Olaf noted that anyone who didn't like fish would likely go hungry in Siberia. It flavors the meat of seal, bear, ducks, and even cow's milk.

While a guest at another Native house, Olaf watched while the daughter of the family filled a wooden bowl with flour, baking powder, and water. After the batter thickened, she rolled it into shape on her bare leg. Each time she lifted the dough Olaf saw a long black streak on it where it had rolled against her leg—leaving that part of the leg comparatively clean, and the only clean spot on her entire body.

He was hungry for bread and ate it anyway.

At one village of about twenty houses, insulating earth was piled atop the roofs, and the windows were all made of ice. Frost formed on the ice windows, and each day they had to be scraped to keep them transparent; in deep cold, a thick ice window is warmer than one of thin glass.

After four months of travel, they reached Irkutsk. From there, Olaf circumscribed the world by riding the Trans-Siberian Railroad to Moscow, traveled on by rail to Europe, booked passage on a ship to New York, and reached Seattle by rail, where, in the spring of 1929, he outfitted the now-repaired *Nanuk.*

Trader Olaf Swenson and his 17-year-old daughter Marion, in 1930. Marion accompanied Swenson on the 1929 voyage of his trading schooner Nanuk. *A writer of his time described Swenson as, "A man with a kind face." He was rugged, tall, and knowledgeable of the Arctic.* LOMEN BROS.

Help from Wien Alaska Airways

While Olaf was off on his great journey, Charles Huntley, the *Elisif's* American radio operator who had joined the Norwegian crew at Seattle, communicated almost daily with the U.S. Army Signal Corps at Nome, and through the Corps to Fairbanks, Seattle, and by commercial radio telegraph systems to the Russian government in Moscow, and even to the families of the *Ellisif's* Norwegian crew in Norway. The *Elisif* had the only radio-telegraph system on the entire Siberian coast that winter. It was owned by the Radio Corporation of America (RCA) and messages were billed at fourteen cents a word.

In February, 1929, Ray S. Polister, a vice-president of Swenson's company aboard the *Elisif*, and a former U.S. Army major, radioed Wien Alaska Airways, a fledgling company at Fairbanks, asking if they would consider flying to the *Elisif* (if the Russian government permitted) to fly the ship's cargo of fur to Alaska. This would allow it to be shipped on by train and ship to New York for sale.

A jittery falling fur market, plus nervousness on Wall Street, gave urgency to the request.

After much consideration, pilot Noel Wien, co-owner of the tiny but growing aviation company, decided to make the flight, thus setting the stage for Alaska's infant commercial aviation industry to demonstrate its strengths, and, and eventually, for a small cadre of heroic pilots to become world-renown.

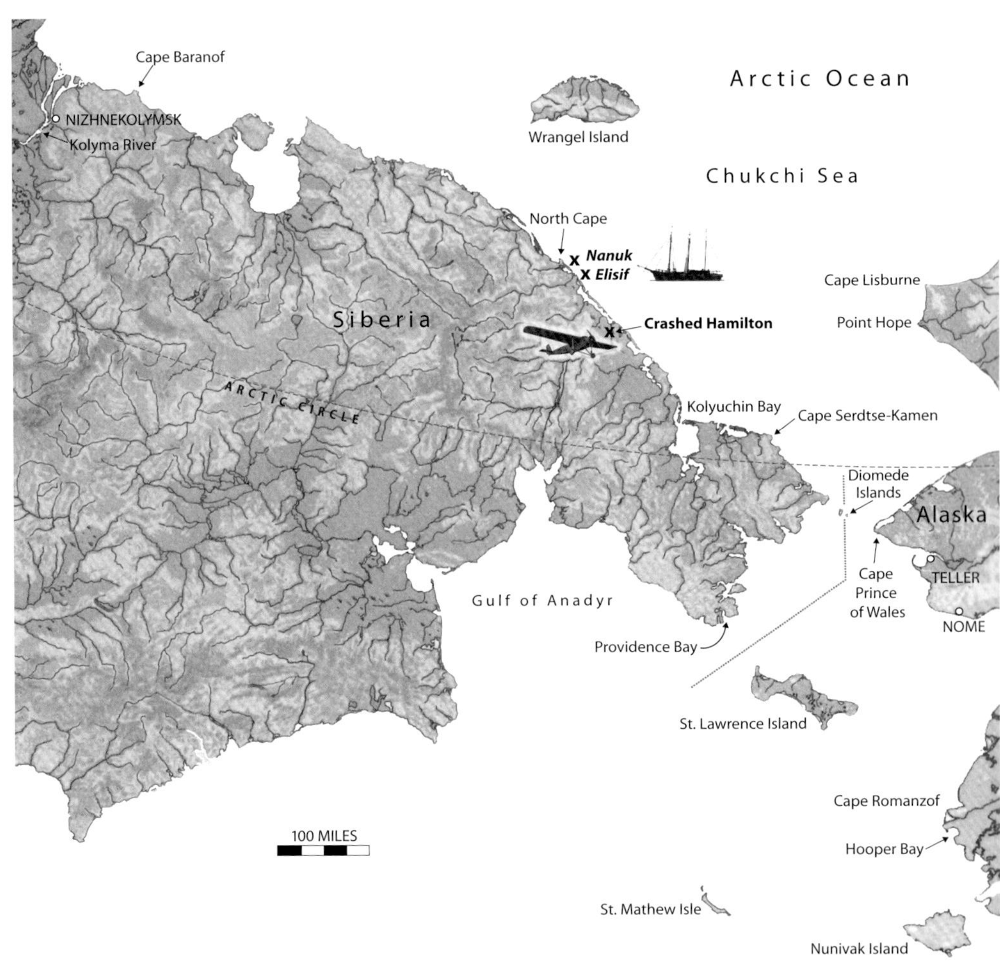

Siberia, where Noel Wien and Ben Eielson flew the Hamilton Metal-plane to retrieve furs from the Elisif *and the* Nanuk.

6 Flight To Siberia

By Noel Wien

[Author] *Following is another gem penned by Noel Wien for the Wien inhouse monthly* Arctic Liner. *His modest description barely acknowledges the extreme hazards of his March, 1929, flights to and from Siberia.*

In February of 1929, my brother, Ralph Wien, was flying the four-place Stinson biplane [the Stinson Detroiter #2], doing charter work out of Fairbanks. I was trying to keep a weekly round trip schedule between Fairbanks and Nome, as well as making charter flights out of both places, with our recently acquired, new, eight-place Hamilton, the finest and largest aircraft in Alaska.

The Trading Schooner *Elisif*

In the first part of that month, we received a wire from the trading schooner *Elisif*. The previous fall it had been frozen fast in the Arctic Ocean, three miles offshore from North Cape, Siberia. The Swenson-Herskovitz Company of New York, owner of the vessel, was the only U.S. company having a concession to trade for furs in Soviet Russia, a regime then unrecognized by the U.S. government.

The vessel had a full load of white fox furs valued at $50 a skin in Siberia, and $100 each in New York, but the market value was shaky. The wire inquired if we would be interested in flying to North Cape (if the Russian government permitted) to bring the fur to Alaska, to then be dispatched by train and boat to New York.

The decision was difficult to make. We had more business than we could handle in Alaska. The risks of flying to Siberia, both weather-wise and otherwise, were great. We had gone heavily in debt for the new equipment [the Hamilton]. During the past year's operation we had not carried insurance of any kind. Life insurance [for aviators] was available in very limited amounts at prohibitive rates; airplane crash insurance for the most part was not available at all.

We [meaning Noel] decided to secure insurance for the flights, but even Lloyd's of London turned us down. We decided to make the flights anyway. This was probably the most risky business decision I made in my thirty years or so of flying in Alaska.

Many wires were sent and received between the schooner, New York fur company offices, Russia, and Fairbanks before final arrangements were completed and permission came from Russia for us to fly to Siberia. We agreed to a price of $4,500 for the first round trip flight of 2,250 miles between Fairbanks and North Cape; subsequent flights were to be made for $4,000 each.

We arrived in Nome from Fairbanks with the Hamilton on February 27. Not until the morning of March 7th did we receive word from the schooner's radio that weather at North Cape was clear, visibility unlimited, and calm. It was -36 F. in Nome, and clear. Winds of thirty to forty miles velocity and temperatures of -30 F. to -40 F. had been prevalent at the *Elisif* for the previous several weeks.

My mechanic for the flight, Calvin "Doc" Cripe, a short but strong fellow, was agreeable, and well liked. He had been a winter stage driver (horses and double-ender sleds) on the Richardson Trail; later he drove trucks over the improved route. He was self-learned as a mechanic and had worked in garages, as well as on airplanes for Owen Meals at Valdez. He had been nicknamed "Doc" because of his ability to repair anything mechanical.

Doc and I removed the seats and other unneeded equipment from the plane, leaving the emergency gear aboard. I took off at 8 a.m. with a 900-pound load of supplies, including a whole hog, and a quarter of beef for the *Elisif.* Our course was over the Alaskan villages of Teller and Wales, across Bering Strait, to East Cape, Siberia.

Crossing Bering Strait

About the time we crossed the International Date Line into tomorrow at mid-Bering Strait, the engine's oil pressure began to increase. We were cruising at 6,000 feet, the outside temperature was -15 F. The oil tank vent, located on the leading edge of the right wing near the cockpit cabin, had frozen up. We had experienced this trouble once before when the cruising altitude temperature had been -10 F. We had fastened a small can over the vent to protect it from the full blast of moving cold air.

By opening a side window, Doc was just able to reach the vent with a knife. He poked and scraped a hole in the closed vent opening. He had to repeat the open-window, cap-sticking procedure every fifteen minutes for the next four hours.

East Cape, Siberia, is about fifty eight miles from Cape Prince of Wales, Alaska's most westerly point. The distance from East Cape to North Cape, Siberia, is roughly 375 miles. There we found the surface covered with heavy, drifted snow. The drifts were well defined from 6,000 feet, and I estimated them to be approximately three to six feet high, fifteen to forty feet apart, and 200 to 300 feet long. This condition I had never observed in Alaska. We realized this was caused by the long prevailing northwest winds and their unobstructed sweep from the north polar region. We could never have landed safely there.

The weather remained clear and visibility unlimited, and Doc's continued operations on the oil vent were successful. We arrived over North Cape six hours and ten minutes after our Nome takeoff. The temperature was forty below zero. Our landing on the ice, one mile offshore and two miles from the ship, was extremely rough, in part due to the frozen oleo[1] struts on the landing gear. Chukchi Eskimos had worked for days with shovels, axes and other implements to make the landing possible. Even so, it seemed the plane would be torn apart as we jumped and bounced on the rough, hard-packed wind-driven snow.

Upon examination after landing, we found the only damage to be small dents in the metal under-covering of the fuselage.

---

1. Hydraulic shock absorbers built into the landing gear.

At North Cape, Wien was greeted by the governor of the village and the crew of the Elisif.*(L to R) Wien, Ray S. Polister the Swenson-Herskovitz fur trading company representative, the governor of North Cape, and Calvin "Doc" Cripe.*

At the *Elisif*

We were taken by dog team the two miles from our plane to the *Elisif.* The drifted snow was so firmly packed from weeks of heavy snowfall and high winds that runners of the sled did not cut into the surface the least bit. When we arrived at the boat we found the drifts over the deck, extending on the leeward side three to four hundred feet. Only the masts of the schooner were visible above the snow.

Everyone was extremely pleased to see us, to receive mail, world news and supplies. We met the governor of the village, and were treated royally aboard that night. While there we received a wire from a Russian government official in Central Siberia, asking if we would make a flight (without charge) to Wrangel Island, 250 miles across moving ice, with food and medicine for a colony of Russians on the island. After our risky, near disastrous landing on the prepared strip, we knew it would be unwise to press our luck further. We sent regrets.

Aboard the *Elisif* were 6,400 arctic fox pelts. They weighed 4,400 pounds, and were worth about $600,000 on the New York market. The plan was for me to haul them to Fairbanks in four flights. In addition, for another possible charter, there were 60,000 sable [marten], and 80,000 squirrel skins at the 500-mile-distant station on the Kolyma River. But first, we were to deliver furs from the *Elisif* to Fairbanks. A machine on the *Elisif* had compressed the fur into bales five or six feet long.

The next morning after a most hearty breakfast, we returned to the plane with sled loads of the baled fur. We packed the cabin full, standing the bales on end, and put about 250 pounds of fur in the fuselage in back of the baggage compartment. We were pressed to take every skin we possibly could.

We gassed up on about seventy gallons of old low grade gasoline which had been stored there for years for use by Russian planes which had not arrived. That, plus gas remaining in the Hamilton's tanks and fifty gallons in cases I had brought from Fairbanks, had to get us home.

Loading 1,625 white fox furs, all bailed—1,100 pounds of it—into the Hamilton Metalplane at North Cape, Siberia, March 5 and 6, 1929. "We packed the cabin full, standing the bales on end. Doc Cripe and I had to get into the plane through the hatch over the pilot's compartment," said Wien.

The temperature hovered around -40 F. The warming of the engine and the landing gear oleo shock absorbers proceeded slowly. We had to shift our one fire pot back and forth between the engine and shock struts.

The Hazardous Take Off

Finally, we were ready and I attempted to take-off. Even after running the engine for a long warm-up, it sputtered and spit from the old low-grade Russian gas. With the throttle wide open I could not raise the heavily loaded tail. We taxied back to unload the furs from behind the baggage compartment. R.S. Polister, the Swenson-Herskovitz representative, tried hard to have us make further attempts to get off with the overload, but I knew it was useless. At this point, the way the engine was acting in the -40 F. temperature and the poor gas, I was not sure we would get off at all. With a lighter load and warmer engine, on the second attempt we just cleared ten-feet-high jagged ice at the edge of the short runway strip.

If we had experienced even a minor mishap then, it is hard to estimate how long we would have been there, and our Alaska business, which we had built up through years of mighty hard pioneer flying, would have been ruined. Both Doc and I breathed sighs of relief and thankfulness after we were safely airborne. Both of us, without speaking of it, had considered the possibility of our being stranded on the *Elisif*, along with the furs.

The weather was again clear and almost calm. Although Doc had to continue his periodic operation on the oil vent, and though the engine ran rough on the poor gas, it did continue to run, and we arrived in Nome in six hours. We completed the flight to Fairbanks the next day.

Successful completion of this flight meant that the first flight between Asia and North America and the first round trip between the two continents had been made. It was noted in many of the papers of the United States.

On the front page of the *New York Times* was a report on the trip, saying that the pilot had carried not only paper credentials authorizing him to make the flight, but also a gun, which "in case of emergency would have been of more value to him."

Noel Wien with his pride and joy, the modern (for 1929) Hamilton Metalplane (not at North Cape).

There was no gun.

As a result of the *Times'* write-up, Wien Airlines received a telegram from the Soviets: "Please stop sending your planes to North Cape until you receive our additional permission."

That permission was never given. However, by successfully making these flights, and retrieving a load of valuable furs, Wien set the stage for one of the great aviation dramas of the time.

[Author] *Ironically, in an age when airplanes were stretching their legs on great flights, and records were beginning to be kept of "first flights" to or between various remote geographical locations around the world, Noel Wien's March, 1929, first flights between North America and Asia, received scant attention. It was casually mentioned in newspaper articles reporting on the retrieval of "a million dollars worth of fur." Modest Noel made little of the exploit, which was unquestionably extremely hazardous.*

At the time, any pilot who had announced that he was planning to make the first ever, record-setting, round-trip flights across the storm-blasted and frozen Bering Sea from North America to Asia (Alaska to Siberia), would probably have received national, if not international attention, and the flights would have been entered in the record books.

Book Two

Eielson's Exploration Flights

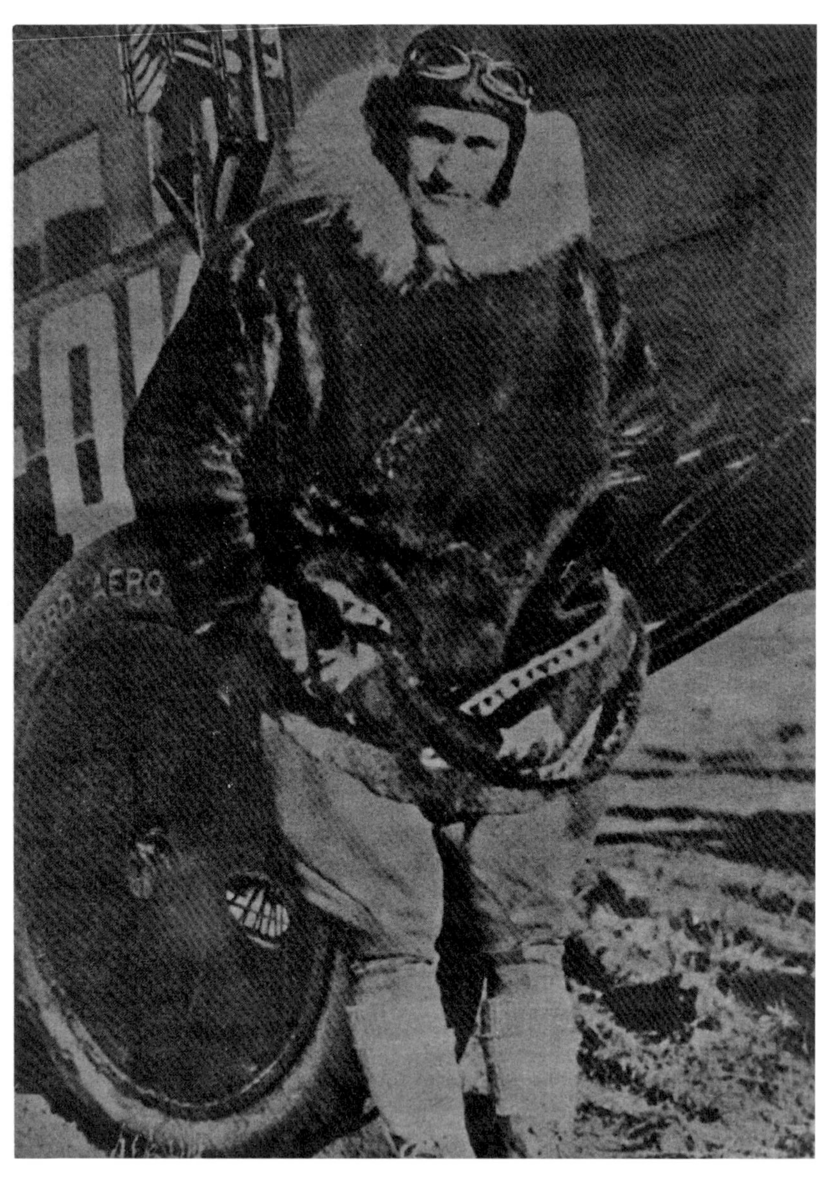

Ben Eielson leans against a wheel of the single-engine Fokker, Alaskan, *brought to Alaska in 1926 by explorer Hubert Wilkins. With Wilkins navigating, Eielson flew this plane on the first flight ever from Fairbanks to Barrow. Today the* Alaskan, *minus its wing, is displayed at the Hatton-Eielson museum in North Dakota.*

7

Eielson and Wilkins' 1926 Flights

George Hubert Wilkins, a self-taught farm boy from the Australian outback, was obsessed by the challenge of exploring unknown lands and little-known places. He is probably best remembered for the early use of airplanes in polar research. Some consider him to have been one of the most successful explorers in history. In his lifetime he was involved in fourteen expeditions, some that lasted for years. He also participated in several record-breaking flights.

Wilkins was an experienced arctic explorer when he chose Ben Eielson to be his pilot for a new polar expedition. He had been a member of the Vilhjalmur Stefansson Canadian Arctic Expedition of 1913–18; was a member of the British Imperial Antarctic expedition of 1919-20, and accompanied Ernest Schackleton's Antarctic expedition of 1921-22.

Born in 1888, he learned to fly in 1910–12. He was 25 when he joined Stefansson's Canadian Arctic expedition, and what he learned in the five years on that odyssey had a great influence on his later career as an explorer.

That expedition was badly organized by Stefansson; food and equipment was loaded helter skelter on the *Karluk*, the expedition's primary ship. Sometimes days-long searches for needed items failed. Most of the personnel Stefansson chose had no arctic experience, and there was much internal dissension. The frail *Karluk,* the ship Stefansson purchased for the expedition, was not suitable for use in the arctic. Wilkins was one of the five expedition members who departed the *Karluk* with Stefansson to hunt caribou in Alaska. At the time, the *Karluk* was trapped in ice that eventually carried her almost to Siberia.

After the debacle of the 1913–1918 Stefansson expedition, whenever Wilkins organized expeditions as he did for much of the rest of his life, he was careful to select qualified people, and went to great lengths to organize the food and equipment that accompanied an expedition.

Crocker Land, or Keenan Land?

Now, in 1926, Wilkins wanted to explore by airplane a million-square- mile region of the Arctic north of Barrow, Alaska, an area no explorer had ever reached. Others, including Stefansson, MacMillan, Nansen, and Storkerson, had reached the fringe of the area with dog teams and sleds, but none of these tied-to-the-ground adventurers had come anywhere near the center.

Wilkins wanted to keep his venture simple, with one airplane and one pilot, in addition to himself. But at Detroit, Michigan, where he organized the venture, things got out of his control. His planned polar search became the Detroit Arctic Expedition, sponsored by the Detroit Aviation Society, the American Geographical Society, and the North American Newspaper Alliance.

Wilkins' "one pilot, one plane" quickly ballooned into three airplanes, and a bevy of mechanics, photographers, correspondents, wireless operators, and pilots. The results were predictable.[1]

Was there a hypothetical "Crocker Land", as proposed by Peary, the first man to reach the North Pole? Or was there a "Keenan Land?" Reporter Palmer Hutchinson, covering the expedition for the North American Newsaper Alliance, wrote about a whaler who, fifty years earlier, had claimed to have seen a shoreline in the unexplored region. That phantom land, according to Hutchinson's article, was "Keenan Land."

1 One official observer loaned to the expedition was Major Thomas G. Lanphier, Com. mandant of the First Pursuit Squadron of Selfridge (Army) Field, Michigan. Lanphier later gained a minor place in history when, on July 1, 1927, he became one of only two pilots other than Lindbergh to ever fly the *Spirit of St. Louis*. He flew it for ten minutes in the vicinity of Selfridge field. The other was Lieutenant Philip R. Love, who, on August 8, 1927, piloted the *Spirit* for ten minutes near the airport at Louisville, Kentucky.

The name Lanphier gained more luster during World War II, when Lanphier's son, Captain Thomas G. Lanphier Jr, in April, 1943, flew one of the four P-38 Lockheed Lightning fighter planes that participated in the shooting down of the Mitsubishi bomber (Betty) which carried to his death Admiral Isoroku Yamamoto, the architect of the Jap attack on Pearl Harbor.

The trimotor Fokker Detroiter, *at Fairbanks to support Hubert Wilkins' 1926 Arctic explorations, crashed on its first flight. It was repaired and made one round trip flight Fairbanks-Barrow. Later sold, it became the famed* Southern Cross *that flew across both the Pacific and Atlantic Oceans. Today it rests in a Canberra, Australia, museum.*

Wilkins was determined to find if there was *any* land in this unexplored region.

Eventually, the Detroit Arctic Expedition had two airplanes; a trimotor Fokker, named the *Detroiter*, powered by three reliable air-cooled Wright Whirlwinds, with a maximum range of 2,500 miles; and a single-engine Fokker, the *Alaskan*, powered by a liquid-cooled Liberty engine, with a range of 3,200 miles. Both were high-wing monoplanes. A third plane, built in Detroit, an all-metal Ford Trimotor, was to have been donated by auto maker Henry Ford. It was destroyed by fire before it could be delivered to Wilkins.

That was the first setback.

It was -40 F. February 23, 1926, when the crates holding the two Fokker planes arrived at Fairbanks and were dragged on horse-drawn sleds from Alaska Railroad flat cars to Rickert's Field. A circus-like atmosphere surrounded every move of the expedition; a battery of movie cameras, hand-cranked by shivering news re-

The Fokker Alaskan, *built in Europe to the specifications of explorer Hubert Wilkins. In 1926 it was flown three round trips Fairbanks to Barrow by Eielson and Wilkins, as well as 125 miles over the Arctic Ocean north of Barrow as Wilkins searched for undiscovered lands. It used a 400 hp, 12 cylinder, liquid-cooled Liberty engine.*

corders even followed the horses hauling the planes to the field.

While in Fairbanks, preparing the airplanes for the expedition, Ben Eielson gave a talk to the faculty and student body of the nearby University of Alaska, outlining Wilkins' goals for the expedition. They were 1. To explore the million square miles of unknown area; 2. To claim for the United States any land that might be found; 3. If land was found, to establish bases for exploration and compilation of scientific data; 4. To demonstrate feasibility of a commercial route over the top of the world; 5. To reach the North Pole by air; and 6. To fly to Spitzbergen over the top of the world.

Death of Palmer Hutchinson

On March 11, Miss Genevieve Parker, winner of a local women's sled dog race, in white parka and mukluks, broke bottles of gasoline on the propellers of the newly assembled two ships. The naming ceremony was observed by many local residents. It had been planned by *Detroit News* reporter Palmer Hutchinson.

He was enthusiastic over the expedition, and he had been busy filing news dispatches.

The ceremony over, the crowd dispersed. With fewer people around and fewer distractions, Wilkins decided to test fly the trimotor *Detroiter*. He sent word of this to Hutchinson, who promptly arrived.

Wilkins chose Lanphier as pilot for the test flight; Wilkins was co-pilot. Engines running, the plane moved ahead, but a snowbank blocked one wheel. Palmer Hutchinson enthusiastically rushed to join mechanics who were tramping snow in front of the wheel. He was close to the spinning propeller of the number one engine, which, because of its speed, was essentially invisible. Eielson and others, frightened, yelled at him and rushed to pull him away. Before they could reach him Hutchinson stepped into the arc of the spinning propeller, and was killed instantly.

The test flight was canceled and a pall settled on the expedition. Hutchinson had been popular.

That was the second setback.

Eielson Crashes the *Alaskan*

For the following week, runway snow-clearing took up much time—both planes were on wheels, not skis, and taking off and landing with snow on the ground would have made plane handling difficult.

On March 15, Eielson was the pilot, and Wilkins the co-pilot of the dual-control *Alaskan* single-engine Fokker for a test flight. They took turns at the controls, and circled for about forty minutes. The plane handled well. Lined up with the runway, Eielson brought her in for a landing. While still about 200 feet in the air, he pulled the throttle back, as he was accustomed to doing with

Ben Eielson and the Fokker monoplane Alaskan *on March 15, 1926, a few minutes after he made a crash landing with it at Fairbanks. The box-like wing construction of this plane is visible next to Eielson.*

the lighter planes he had always flown. The ship stalled and hit the ground fifty feet short of the runway. The landing gear collapsed, one tip of the metal prop was curled, and the under-cowling was dented. The plane skidded a short way. Neither man was hurt.

That was the third setback.

In retrospect, one has to wonder why Noel Wien, who witnessed the two accidents, was not asked to test fly the Fokker. He had flown similar aircraft for more than 100 hours, landing on shorter runways. It might have changed history.

The landing gear was repaired, a replacement 400-horsepower Liberty engine was installed, and a wooden prop replaced the ruined metal one.

Nothing bruises a pilot's ego more than breaking an airplane. It was Ben Eielson's first serious airplane crash (not counting the wing-wrapping occasion with the Jenny). He was deeply mortified.

Lanphier Crashes the *Detroiter*

A few days later with Major Lanphier at the controls and Wilkins co-pilot beside him, the three-engine *Detroiter* took to the air. During its take off run the big Fokker veered left on the rough field, and Lanphier corrected. The plane barely staggered into the air over a high snow berm.

After circling for a time, Lanphier lined up with the runway for a landing. When he was about 100 feet above the ground, he cut power to the two outboard engines, planning to glide to the ground with power from the center engine.

The big plane stalled, as had the *Alaskan* earlier with Eielson. It slammed into the ground, one landing gear collapsed. Neither man was injured, but the airplane was badly damaged. The motor mount for the center engine was not repairable, and its cowling was wrecked; the propeller for the right hand engine was badly bent, with possibly a bent crankshaft.

That was the fourth setback.

Neither Eielson or Lanphier had flown such large airplanes as the two Fokkers. Lanphier had no experience with a multi-engine ship. Both learned a bitter lesson. The big Fokkers' flying characteristics were markedly different from the light planes with which they were experienced. Today, before a pilot could fly a new type of airplane, a check ride with a pilot familiar with it would be normal.

When Lindbergh allowed Lanphier to fly the *Spirit of St. Louis* was he aware that the Major had broken the *Detroiter* while trying to land?

The Overland Party

Wilkins planned to make his exploratory flights over the frozen Arctic Ocean from Barrow, Alaska's farthest north village. The plans for this involved an overland party that would haul to Barrow the fuel necessary for such flights. To do this, three Snow Motors were purchased. These hybrid machines were standard Fordson tractors which had been modified by removing the wheels, and substituting two three-feet-diameter pointed cylinders, one mounted on each side of the tractor. Wrapped around each cylinder, full length, was a four-inch-high spiral or screw. When the cylinders revolved, the screws were supposed to pull the machine through the snow.

Explorer Hubert Wilkins at Fairbanks in 1926 standing next to a wheel of the trimotor Fokker Detroiter.

The Snow Motors were to tow double-ender sleds, of which ten were ordered at $700 each from Cal M. Brosius at Seward. Fifteen tons of supplies, mostly aviation gas and oil, plus fuel for the snow motors, was assembled for the overland haul from Nenana to Barrow.

Two of the machines (the third was left at Nenana as a spare) departed Nenana February 11. They traveled a mere seventy-five miles in nineteen days, suffering numerous breakdowns and delays. They had to be dug repeatedly out of deep snow. They had worked in wet, heavy snow in Michigan, but they hardly worked at all in the dry, light, and loose snow of Alaska's Interior. The attempt to reach Barrow with them ended.

That was the fifth setback, and it altered major plans for the expedition.

Eielson Flies the *Alaskan* to Barrow

Eielson redeemed himself on March 31 when, with Wilkins navigating, he flew the *Alaskan* from Fairbanks to Barrow with 3,000 pounds of fuel and other cargo.

Alaska was still a land of snow. The heavily-loaded Fokker easily climbed over the low hills north of Fairbanks and crossed the scattered spruce-birch forests to the eighty-mile-distant still-frozen Yukon River. Airspeed was ninety-five miles per hour, with a tail wind of forty-five mph.

With 160 miles behind them, they reached the southern front of the rugged, glaciated Brooks Range that swoops in a gentle arc from east to west across Alaska, from Canada to the Arctic Ocean.

The sun was dazzling on the 7,000 to 8,000-feet-high snowy peaks ahead. Mount Doonerak, close to their route, is 7,457 feet. The U.S. Geological Survey map Wilkins used for navigating showed the highest peak they had to cross was 5,000 feet.

As they neared these imposing giants, Ben nursed the big plane to 9,000 feet. Steam from the radiator of the hard-working liquid-cooled engine smeared the windshield, and they had to peer from side windows to see out.

Their plan was to follow up the John River from Bettles and cross the Brooks Range through Anaktuvuk Pass (Eskimo for "place of many caribou droppings").

After about eighty miles of peering at some of the most spectacular snow-blanketed mountain scenery in North America, they were relieved to find the high rugged peaks ended abruptly on the north side at the arctic foothills. They now droned above snow-covered, rolling hills and uplands that soon gave way to the gentle plains of the North Slope, famed today for the great Prudhoe Bay oil fields. This vast treeless region is a northern desert, underlain by permafrost (permanently frozen ground). Since 1977, billions

One of the three Snow Motors purchased by Wilkins for his 1926 Detroit Arctic Expedition. The machines were modified Fordson tractors, wheels removed and replaced with rotating screw-equipped cylinders. They were supposed to tow sleds from Nenana to Barrow with fifteen tons of supplies, mostly aviation gas and oil.

They didn't work well in the light, dry snow of Alaska. Two of the machines covered seventy-five miles in nineteen days, had numerous breakdowns and delays, and the effort was abandoned. This unit is on display at the Pioneer Air Museum in Fairbanks. Author

of barrels (17% of U.S. production) of the black gold have poured through the four-foot-diameter 800-mile-long pipeline from the North Slope to Alaska's port of Valdez.

They breathed a sigh of relief at putting the great mountains behind them, as they started across the last 250 miles to Barrow.

While crossing the Arctic Plains, solid clouds lay below as they flew at 5,000 feet. A strong tailwind continued to push them. Clouds ended at the shoreline of the Arctic Ocean. All continued white below. Soon, Wilkins recognized the character of the ice, and realized they had left land and were now over the pack ice of the Arctic Ocean. During the 1920s and decades afterward, until recently, offshore pack ice existed here for at least ten months of the year.

First Flight Over an Unknown Region

Wilkins, ever the explorer, decided to fly on. After an hour, he told Eielson, "You can now see 100 miles farther north in this area than any other man has seen until today. We're 100 miles beyond land over the Arctic Ocean. What do you think about flying north for another half hour?"

"Whatever you think best," Eielson answered. They continued to fly over the frozen sea.

On their return to land they encountered heavy falling snow, and in limited visibility their search for Barrow seemed hopeless. Finally, they saw a vertical bluff marking the edge of the sea, and followed it to Barrow, where Eielson made a smooth landing on the frozen Barrow Lagoon.

That was the only flight the expedition made over the ice pack in 1926. At the time, this flight by Hubert Wilkins and Ben Eielson may have been the longest nonstop flight made in the Arctic.

Theirs was the first flight across the rugged Brooks Range, and the first airplane to span the North Slope to land at the village of Barrow, Alaska's northernmost village, 500 miles from Fairbanks.

That season Wilkins made three more freighting trips from Fairbanks to Barrow, two of them with Eielson as pilot of the *Alaskan.*

The *Alaskan* Drops a Wing

On May 6, Eielson tried to take off with the loaded *Alaskan* for another flight to Barrow, but the plane couldn't get off; it overran the runway. Some cargo was removed, and Eielson again sent the big ship down the runway. He had almost reached flying speed when the plane struck a hump, followed by a hollow. The right wing whipped up and down and broke off next to the fuselage. The right landing gear folded, dropping the plane's fuselage to the ground and ruining the much-repaired wood propeller.

That was the sixth setback.

Wilkins was standing behind Eielson. Thirty loose cases of gasoline skidded along the cabin floor, and some hit him. Fuel gurgled from the tank of the dropped right wing. Fearful of fire, both men were trapped until the gas cases were removed by an anxious crew.

The wonder was that the obviously flawed wing didn't fail in flight, given the three loads the plane had flown to Barrow, and the air turbulence it had encountered. In that event, both Eielson and Wilkins would likely have been killed.

The wing could not be repaired; the *Alaskan* was through for the season. However, it would come to life the following year as a hybrid, mounted with the longer wing transferred to it from the tri-motor *Detroiter.*

The *Detroiter* Flies to Barrow

On May 8, the trimotor *Detroiter,* now repaired, with Major Lanphier at the controls, flew 700 gallons of gas and 45 gallons of oil to Barrow. The snow was gone at Fairbanks, but, looking to the future, a new set of nine-feet-long and twenty-inch-wide skis were being built for it by mechanic Jim Hutchison. These were mounted on the *Alaskan* the following year.

Byrd and Amundson Reach the Pole

While Wilkins waited for fog to clear at Barrow so the *Detroiter* could return to Fairbanks, on May 9, U.S. Navy Commander Richard E. Byrd, with Floyd Bennett as pilot, flew in the trimotored Fokker F.VII-3m *Josephine Ford* from Kings Bay, Spitzbergen, to the North Pole and back. (In recent years Byrd's claim of having reached the North Pole has been challenged, by, among others Bernt Balchen, the pilot who in 1929 flew Byrd over the South Pole. The air speed of the *Josephine Ford*, the reported elapsed time, and the distance flown do not jibe).

Two days later, on May 11, Roald Amundson, Lincoln Ellsworth, and Umberto Nobile departed Spitzbergen aboard the Italian-built dirigible *Norge* (meaning Norway), also bound for the North Pole. They dropped flags there and remained for an hour, afterward heading for Alaska. The *Norge* was sighted flying near Barrow on May 13. It landed on the Seward Peninsula, at Teller, Alaska, at 8:30 p.m. that same day.

Spitzbergen is 400 miles north of Norway, on the far side of the Arctic from North America, and opposite the area Wilkins wanted to explore.

A Jinxed Expedition?

The 1926 Detroit Arctic Expedition seemed jinxed; a donated Ford Trimotor burned before delivery; both Fokkers crashed on their initial flights; popular newsman Palmer Hutchinson was killed by the propeller of the *Detroiter*; an expensive overland freighting attempt failed; only one exploration flight was made over the Arctic Ocean, although multiple flights were the goal.

On the other hand, Wilkins, Ben Eielson, and Major Lanphier, had demonstrated that airplanes could cross the vast Brooks Range to Barrow. Ben Eielson, with the *Alaskan*, flew three successful round trips; Major Lanphier flew the round trip once with the *Detroiter.* Wilkins had navigated all of these flights. One flight of 125 miles over the Arctic Ocean with Eielson at the controls of the *Alaskan* had revealed no land—a negative finding, but still informative.

By June, 1926, the window of opportunity for exploration flights over the Arctic Ocean had passed. The occurrence of fog along the Arctic coast greatly increases when open water begins to appear in May or June, and this condition continues until early October. (These dates may not be valid today; great losses of Arctic Ocean sea ice have taken place since the events recorded here). An early breakup had left Fairbanks landing fields too soggy for heavily loaded planes to take off. For these reasons, in early June Wilkins called it quits for the year.

However, he hadn't given up on his dream of flying over the last great unexplored region of the Arctic. He was already planning to return to Barrow in 1927 with the unchanged goal of flying over the unknown region north of Barrow. He had a substantial cache of aviation gas at Barrow as a start. George Hubert Wilkins didn't give up easily; success was to be his. Ben Eielson's rising star was linked to that of Wilkins.

8 Eielson and Wilkins' 1927 Flights

DURING THE EARLY WINTER of 1926–27, Eielson became First Pilot for Florida Airways, where he flew a tri-motor plane with passengers and mail. He returned to Alaska in February to prepare for Wilkins' 1927 expedition.

The failures of his 1926 expedition seemingly didn't discourage Wilkins. He was back in Alaska in 1927 with another expedition, his goal unchanged; he was still determined to explore a vast unknown region of the Arctic from the air.

Ben Eielson was the only member of the 1926 crew with Wilkins this year; he was Chief Pilot; Alger Graham, who had flown a Stinson for the Waco Oil Company of Detroit, was his Second Pilot. Two Stinson Standards, *Stinson Detroit News Number 1*, and *Stinson Detroit News Number 2,* were to do the flying. With a 220 hp Wright Whirlwind engine the new five-place (pilot and four passengers) cabin biplane was capable of taking off in 1,200 feet with a 2,000-pound load.

OVER THE ARCTIC ICE AGAIN

On March 25, 1927, in seven hours, both Stinsons were flown nonstop, Fairbanks to Barrow. Eielson flew one, and Graham the other. A.M. Smith, the *Detroit News* correspondent (the *Detroit News* supported the expedition financially) had to be left behind, for both Stinsons were full. Arrangements were made to have him flown to Barrow by Joe Crosson in the Fairbanks Airplane

For his 1927 Detroit News-Wilkins Expedition Hubert Wilkins purchased two identical Stinson four-passenger airplanes, designating them as Stinson Detroiter No. 1 *and* No. 2. *Number 1 was lost when it crash-landed out of fuel when Eielson and Wilkins flew it over the Arctic Ocean searching for new land. The two men left the wreck and walked nearly 100 miles back to Alaska over shifting arctic ice.*

Corporation's Hisso-powered open cockpit Super Swallow C2375 (See chapter 15 for the story of Crosson's flight).

On March 29, with Eielson at the controls and Wilkins navigating, the heavily loaded (with fuel, emergency gear, and water-depth-sounding equipment) *Detroit News Number 1* left Barrow and flew north over the unexplored region of Wilkins' dreams. The temperature was -42 F. at Barrow.

Four hours and forty-five minutes into the flight, 450 miles from Barrow with rough ice below, the engine started to miss and backfire. Eielson managed to keep it running for another thirty miles to smooth ice where he carefully planted the skis of the biplane.

He was the first pilot known to land an airplane on the arctic pack ice.

Eielson probed to find the problem with the engine while Wilkins, anxious to determine water depth (one of the expedition's goals), chipped two holes in the three-feet-thick ice. Ben left the engine long enough to detonate a charge in one of the holes. Wilkins measured the time it took for an echo to return from the bottom. The depth exceeded 5000 meters (3 miles) the deepest place found in the Arctic Ocean to that date, an indication there was little likelihood of nearby land.

Both men worked on the engine for another two hours. In that time, Ben froze four of his fingers in the -32 F. temperature and twelve-mile-an-hour wind. The engine re-started, but it wouldn't turn its full rpms.

Despite having less than full power, Ben pulled the little biplane into the air, but after ten minutes of flight, the engine again started to miss and knock, and Eielson found another smooth ice floe and made a second safe landing. They spent an hour working on the ignition system, and again took to the air, this time with the engine running smoothly.

They still had eight hours of fuel, but they flew into a forty-mile-an hour crosswind and had to crab into it as they headed toward Barrow. At 3,500 feet the wind was enough against them to cause the loss of twenty miles an hour in ground speed. By 6:00 p.m. they were still in a forty-mile-an-hour wind at 5,000 feet.

Crash on the Ice

Dark arrived, and they were low on fuel. A little after 9 p.m. the engine quit, out of gas. Ben, on instruments, glided through complete darkness toward whatever was below, knowing it was mostly jagged and broken ice. The plane, near the end of its silent last-ever glide, rocked in turbulence, and flew into ground-drifting snow. The landing gear hit an ice ridge, and the plane rolled to one side. It slammed into snow drifts with the left wing and skis hitting at the same moment. They skidded briefly and stopped.

A flashlight showed both ski pedestals were wrecked, and the skis turned on their sides. It was dark, with blinding blowing snow, and they couldn't see their surroundings. Exhausted, they returned to the cabin and climbed into sleeping bags. They fell asleep to the sound of wind and snow hissing on the fuselage of the wrecked plane.

Daylight revealed their broken airplane lying in a fifteen-by-thirty-yard flat area surrounded by jagged pressure ridges, a spot they never would have considered landing on in daylight. To have landed blindly in a level area with a minor crash was miraculous.

Some of the objectives of the expedition were reached on their 550-mile northbound flight. Without seeing land, they had flown to latitude 77 degrees 45 seconds north, 175 degrees longitude west.

They had learned that practical landing sites on the ice were plentiful proving that planes could be safely used for exploring the arctic (some arctic experts had ridiculed the idea of landing an airplane on the ice pack, claiming it was impossible). Wilkins' sounding of water depth helped define the edge of the continental shelf.

The storm continued next day. Wilkins chopped a hole through the ice and determined the floe they were on was drifting northeast at five or six miles an hour. Late that day, two shots of a momentary appearance of the sun told him they were 100 miles northeast of Barrow. Daily revaluations of their position indicated the floe they were on drifted northeast, then east, and finally southeast, covering about 200 miles within five days.

Offshore, north of Barrow, the current in the Arctic Ocean is usually to the west at about one mile per day. At the time Eielson and Wilkins were stranded, huge ice floes from seven-to-thirteen-feet-thick covered enormous areas of the Arctic Ocean in winter, and well into March. These floes floated free of the ground except in a few shallow areas, so they drifted with the prevailing winds and currents. These generally pushed the ice from the northern coast of Siberia toward the northeastern corner of Greenland, straight across the North Pole. There are numerous local gyres, where tidal currents circle.

The sea ice of the Beaufort and Chukchi Sea is never a solid mass. Some linear leads or openings in the ice, and polynyas—large non-linear openings in the ice—are always present.

The Long Walk to Land

Eielson's and Wilkins' lives were now dependent upon Wilkins' knowledge of arctic survival which he had gained while on expedition with Stefansson. Stefansson had learned to live on the ice from Eskimos.

The storm lasted several days. Finally, on a clear day with the temperature -60 F., they started for eighty-mile distant Beechey Point and the trading post there. They made two sleds from airplane parts on which they dragged sleeping bags and emergency food. At times their floe traveled faster than they did, and in the opposite direction of their travel; thus after a day of very difficult travel, they could lose ground.

They struggled through hip-deep snow, and crawled across

pressure ridges, some up to twenty feet high. Eielson's frost-bitten and frozen fingers made it impossible for him to grasp anything. Within a few days they abandoned their sleds as being too difficult to drag across the rough and broken ice. Now they each carried eighty pounds of gear on his back. They abandoned one sleeping bag, and at night slept in their parkas while both stuffed their feet and legs into the remaining sleeping bag as they slept. Each night Wilkins built a snow house, which sheltered them from wind, and allowed them to rest. These houses undoubtedly saved their lives.

While crossing a lead covered with new ice, Wilkins broke through. He managed to struggle out. It was -10 F., and his clothing quickly froze. He ran to a nearby pressure ridge for shelter from wind where, with loose snow, he blotted much of the water from his outer clothing. With Eielson's help, he replaced sodden boots and socks with dry ones. After squeezing all the water possible from his clothing, he wore it wet, for he had no extras. His body heat dried it in two days.

There is one plus to wearing frozen clothes; although heavier and stiff, they are more windproof.

For thirteen days they struggled across the ice, frequently having to detour around open water and new ice too thin to support them. They sometimes had to crawl where ice was so rough they couldn't walk. At night they spent an hour or more building a snow house, and knocked frost from their clothing before bedding down. They slept with their woolen socks inside their clothing to dry them.

They reached the trading post at Beechey Point operated by Tony Edwardson on April 15. An Eskimo with a dog team left for 120-mile-distant Barrow with a letter from Wilkins to Alger Graham, asking him to retrieve them with the *Detroit News No. 2.*

At Barrow Dr. Newhall amputated two joints of Eielson's right hand little finger which had turned black. His remaining fingers, mostly frost-bitten and not frozen through, eventually healed.

Eielson Heals; Wilkins Flies

Graham flew Wilkins and Eielson to Fairbanks. While Eielson healed his frostbitten hands, and recovered from the ice-crossing

Noel Wien flew one of the Stinson Detroiters for Wilkins for about two weeks at Fairbanks during the organizing of Wilkins' 1927 arctic expedition. Shown here with one of the two Detroiters in a rare photo are (l to r) Noel Wien, Hubert Wilkins, and Ben Eielson.

ordeal, Wilkins and Graham left in the *Detroit News No. 2* to make another flight from Barrow, this time to the east. They flew 400 miles over ice and occasional water without seeing any land, although clouds somewhat restricted visibility.

In June, 1927, with the work ended for the season, Wilkins disassembled and stored the *Detroit News No. 2* at Fairbanks. He planned to return in 1928 to again fly it into the Arctic. Later, during winter 1927-28, after seeing an airplane he thought more appropriate for his plans, he sold the airplane to Noel and Ralph Wien (See Chapter 3).

That summer and winter, while waiting for Wilkins' 1928 arctic expedition, Eielson worked for the new federal Bureau of Aeronautics, which had started to license airplanes, tested new airplanes before allowing them to be marketed, checked out new airfields, organized a standard traffic pattern around airfields, and investigated aircraft accidents. The job was wide-ranging, and on official business he flew across much of the United States.

9 The Flight to Spitzbergen

Wilkins Buys A New Plane

Robert Peary, who had reached the North Pole, had years of experience in the Arctic. He believed land would be found in a vast area east of Barrow. He thought he had seen such land in the distance in an unexplored region. The goal of the Wilkins 1928 expedition was to explore this unknown region from the air. Most arctic experts were skeptical of the value of learning what was in the region. The great Roald Amundsen questioned the wisdom of attempting to fly across this region, suggesting it was beyond the ability of current airplanes.

While in San Francisco during the winter of 1927–28, from his hotel room Wilkins caught a glimpse of a streamlined monoplane as it flew over. The airplane didn't have lift struts supporting the wings. There were no wires or other drag-inducing structures. Its very appearance shouted speed and efficiency. Excited by the airplane, he tracked it to an Oakland, California, airfield.

There he found the prototype Lockheed Vega ("veega"), called the *Golden Eagle*. It was being prepared to compete in the Dole race, from Oakland, California, to Hawaii. Sadly, it was never seen after it left Oakland August 16, 1927, bound for Hawaii.

Wilkins went to Los Angeles where he found the Lockheed factory in a rented workshop with a second Vega nearly completed, and a third one started. The builder was Allen Loughheed, the designer Jack Northrup. The design was unproven, and the airplane was not in quantity production.

Nevertheless, when Loughheed and Northrup showed him performance data, Wilkins ordered one, and Lougheed custom-built it for his 1928 expedition.[1]

Over the next few years various Vegas were flown by Amelia Earhart (New York to Ireland, and Hawaii to California), Frank Hawks (a Los Angeles to New York record), and many other celebrated pilots.

The most famous Lockheed Vega was the *Winnie Mae* owned by Wiley Post. It broke speed and distance marks of all types. Post flew it around the world twice (in 1931 and 1933). In later years he piloted it on high altitude research and he is credited with finding the jet stream with it. Today the *Winnie Mae* is on display at the Steven F. Udvar-Hazy Center Smithsonian Museum, near Washington, D. C.

Vegas were popular in Alaska, where from 1929 through the 1930s, they were flown by various bush airlines, both in the Southeastern panhandle, and elsewhere. In Southeastern Alaska they were commonly mounted on floats. The fabric covering the fuselage and wings on early models wasn't waterproof, and the wood could soak up as much as 100 pounds of water during a flying season in the wet world of Alaska's panhandle. This defect was later corrected.

Vegas had a couple of interesting traits that kept pilots and passengers aware they were flying in an all-wood airplane. If over-primed, sometimes gasoline leaked from the updraft carburetor and dripped from the air scoop below the cowling. Occasionally it caught fire from the nearby exhaust stack, and flames shot around the front of the fuselage. Pilots in the know used a lot of throttle to blow the fire out.

In cold weather and in air turbulence, the three-sixteenth-inch bonded plywood that formed the fuselage snapped and popped

1. The Lockheed Vega (Loughheed altered spelling of his name for simplicity) was a far advanced design for 1927. The fuselage framework was built up of two laminated plywood shells formed to shape in a mold, and then assembled over circular wood formers that were held in line by a few wood stringers (monocoque construction). The skin carried the major portion of fuselage stresses. After all cutouts (windows, gas tank opening, etc.) were made the fuselage was fabric-covered for increased strength and better finish. The plywood wings were also fabric-covered for the same reason.

The Lockheed Vega X3903 (X for experimental) in which, on April 15, 1928, Eielson and Wilkins flew twenty hours and twenty minutes from Barrow, Alaska, 2,200 miles across the frozen Arctic Ocean to Spitzbergen, Norway. It was painted a deep orange (highly visible if parked on ice), with "Detroit News – Wilkins Arctic Expedition" in contrasting blue letters on each side of the cockpit. Wilkins is on the left in this photo, Eielson on the right.

loudly—sounds pilots and passengers preferred not to hear when airborne.

When taxiing at high speed in rough water, or when aloft in turbulent air, when the airplane flexed, passengers could tell how severe it was by watching the upholstery on the back of the cockpit bulkhead; it tightened or slacked as the airplane flexed.

Vega X3903 (its federal registration number, the X meaning experimental) built for Wilkins was the third Vega built. It had a 220-hp Wright Whirlwind engine, extra gas tanks in the fuselage, and two extra wing tanks. Theoretically, it had a range of 3,300 miles. Windows were built into the floor for convenience in making drift observations while aloft, and for noting ground conditions. There were two windows on each side of the cabin. A hatch atop the fuselage allowed a navigator to make celestial observations. The pilot had good visibility through the windshield, and there was a window on each side of the cockpit.

The airplane was painted a deep orange, with "Detroit News

Noel Wien had a fine camera and took many great photos of Alaska's early planes and aviators. Here is a superb photo he took of Wilkins' Lockheed Vega on March 5, 1928, as Ben was preparing to land at Rickard's Field at Fairbanks. Eielson and Wilkins flew the Vega non-stop to Barrow on March 20, in preparation for their great flight across the top of the world.

– Wilkins Arctic Expedition" printed on its side in large, contrasting, blue letters.

The Old Planes are Sold and Given Away

To pay for the Vega, Wilkins sold the tri-motored Fokker *Detroiter*, and the *Stinson Detroiter Number 2.* The Wien brothers made good use of the *Stinson Detroiter Number 2* through two winters, flying between Fairbanks and Nome. Wilkins later commented that the sale of the airplane to the Wiens made it possible for him to buy the Vega.

The Fokker *Alaskan* today resides in the Hatton-Eielson Museum, in Hatton, North Dakota, Ben Eielson's home town. Sir Hubert Wilkins gave the airplane to Ole Eielson, Ben's father, in 1930, after Ben's death. It is a Fokker F-VIIa, built in 1926 specifically for Wilkins, and was powered with a liquid-cooled 1917 Liberty twelve cylinder 400 hp engine built by the Ford company.

It arrived in North Dakota without its wing and the engine cowl, and is on display in that form.

When last flown by Wilkins in 1927, the *Alaskan* used the borrowed wing of the *Detroiter*, the Fokker Trimotor Wilkins used on his 1926 Arctic expedition. That wing was probably returned to the *Detroiter* when the *Detroiter* was sold to Australians Charles Kingsford Smith and Charles Ulm, who re-named it the *Southern Cross* and made it famous. They flew it from Oakland, California, to Australia in three jumps, establishing a world record for distance on the 3,138-mile flight from Hawaii to Fiji. Later they flew the big trimotor around the world, which included a flight across the Atlantic from Ireland to Newfoundland. The *Southern Cross* is now in a Canberra, Australia, museum.

Eielson Flies the Vega

Eielson left his job as an inspector for the Bureau of Aeronautics and went to California to fly Wilkins' new Vega. On one test flight, with Wilkins as passenger, he flew to Muroc Dry Lake, where the plane took to the air in 2,700 feet with the load expected for the planned flight across the Arctic. The new Vega cruised at 134 mph, high speed for the time. Later Vega models were equipped with a 450 h.p. Pratt & Whitney Wasp engine, which gave them a maximum speed of 195 mph, and a cruising speed of 170 mph.

The Vega Flies to Spitzbergen

On March 20, with the temperature -24 F., Eielson flew the Vega from Fairbanks to Barrow, with Wilkins navigating from behind the pilot's compartment. It was -48 F. at Barrow when Ben landed on the frozen lagoon there.

Eskimo women sewed fur clothing for the two. A crew of Eskimos, each receiving six dollars a day, shoveled a runway clear of snow. Two attempts to take off failed; each time the plane veered from the runway into soft snow. Finally, on April 15, Ben shoved the throttle of the heavily-loaded Vega wide open and sent it down the narrow, most-recently cleared runway. Included in the load was an extra steel propeller, snowshoes, skis, a rifle and 400 cartridges, fish nets and hooks, sleeping bags, extra arctic clothing.

Snow berms on each side of the runway, piled by shoveling, were higher than the wings. The plane occasionally swerved on the icy surface. Eielson danced on the rudder pedals to keep the plane centered. At times, wingtips came with a foot of the berms. A wing hit on a berm could have ended the expedition.

Speed built. The berms became blurs to the explorers. The plane lifted briefly at seventy miles an hour, touched down momentarily, and finally climbed into the dense, icy air.

Spitzbergen, an island, lies 400 miles north of Norway. Wilkins planned their great circle course to pass a point about 300 miles south of the geographic North Pole. They would cross regions in the Arctic Ocean where Peary and others thought they had seen land in the distance.

They had to take into account the magnetic deviation, and because of the narrow convergence of meridians in the Far North, they had to change course fifty times; over one segment of the route they had to change course every five minutes.

They first flew 500 to 800 feet above the jumbled ice and open water. Ben climbed the plane as gasoline was burned off, and it became lighter. At first the weather was clear. After three hours they flew into low cloud cover, and Ben had to occasionally go mostly on instruments.

When they had flown for thirteen hours they flew past great jagged peaks that thrust above the clouds. Wilkins thought they were near Grants Land, near Cape Columbia. There they had a choice; they could continue above the clouds and land on Greenland; or they could head directly for Spitzbergen with the likelihood of running into bad weather.

They chose Spitzbergen. Both men were cold and uncomfortable.

They came to high cumulous clouds, and Ben climbed the plane to 8,000 feet and circled around clouds. They had about four hours of fuel left when, far ahead they saw mountain peaks they thought might be on Spitzbergen. Ben descended. The air became turbulent. They flew near land, which was deep in snow. Ice and water lay below. Ahead, seen through flying snow, was a coastal headland. Visibility was decreasing in snow. They decided to land. Now their fuel was mostly gone.

They flew past a smooth-appearing area on the nearby shore.

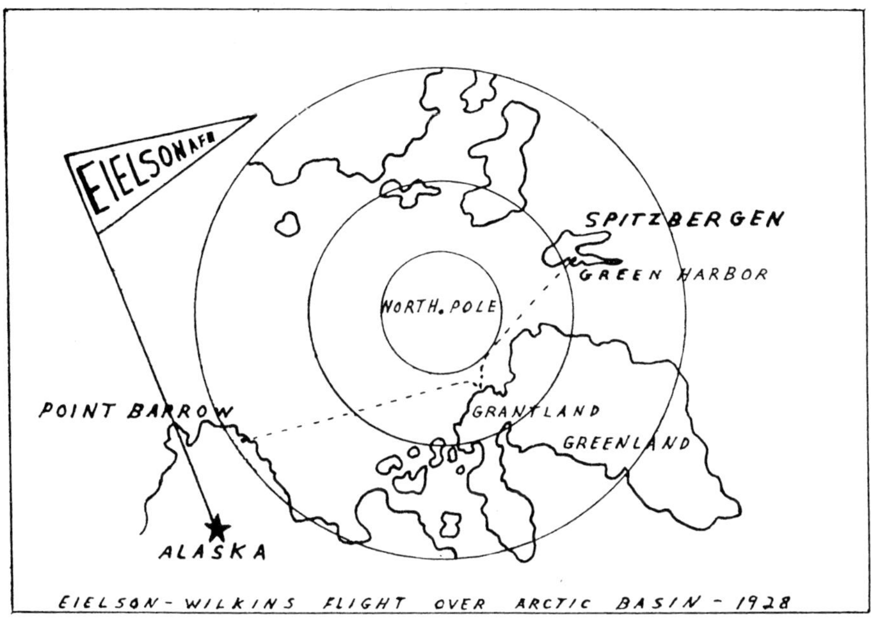

The route followed by Eielson and Wilkins on their great Barrow to Spitzbergen flight. They flew a great circle, and passed about 300 miles from the geographic North Pole. To accomplish this, they changed course fifty times; over one segment of their route they changed course every five minutes.

Ben circled and descended through wind-driven snow. The wood skis of the Vega gently eased into the snow in a perfect landing.

They leaped from the plane, drained the hot engine oil into a five-gallon can, and kicked and stomped snow over the skis to keep the plane anchored.

They slept for a time in the plane as it rocked in a strong wind. Later, Wilkins got a shot of the sun, which put them on the west coast of Spitsbergen.

Next day, snow was knee deep, and still falling. They skied to a small hill half an hour away and found a wooden navigation marker. It was there to guide mariners.

The storm continued for five days. Finally, on April 17, 1928, weather became flyable. They had twenty gallons of gas left.

They shoveled a runway, heated the engine oil with a primus stove, poured it in, and started the engine. Even with full power, the plane wouldn't move. Wilkins had to push and hope he could pull himself in as the plane started moving.

Ben took off without him. Wilkins lay where he had fallen in the deep snow. Aloft, Ben looked back to see his companion still on the ground. He circled and landed. Again Wilkins pushed to get the plane moving. Again he failed to climb aboard, tumbling in the deep snow as the Vega left him behind. Again Ben landed.

This time, to reduce drag, they blocked the tail up with snow. Wilkins used a piece of driftwood, and while remaining part way inside the plane, pushed with the wood with all his strength while Ben used full power. The plane moved slowly at first. It gained speed, jouncing over the uneven ground, and finally lifted off with both men aboard.

As they rounded a headland, five miles across a bay they saw two radio masts and a cluster of houses. They flew across open water and Ben landed near the masts. They were at Green Harbor, Spitzbergen. It was Saturday, April 21, and they had flown twenty-one hours and twenty minutes to cover 2,200 miles from Barrow.

It was an amazing feat. They had covered three times as much territory in one flight as Peary had covered in twenty-three years of exploration.

They found no new land.

Afterward

Four English-speaking Norwegians, officials of the government radio at Green Harbor, arrived on skis and welcomed them. Ben, who had learned some Norwegian as a boy, greeted them in that language.

They were given free use of the government radio (Svalbard Radio), and immediately sent telegrams, Ben to his family, Wilkins to the *Detroit News* which said, "No foxes seen," code for no land seen.

The news of their flight spread, and congratulatory wires arrived from the King of Norway, President Coolidge, the Secretary of State, the National Geographic Society, and explorers Nansen, Stefansson and Amundsen, and many others.

Wilkins sent wires to Lockheed, the Wright Aeronautical Company, Richfield Oil Company, Pennzoil Oil Company (which had provided them with engine oil), and others.

Their feat made headlines around the world. Wilkins and Ben received many wired invitations to visit various European capitals. Amundsen invited them to visit him at his home in Norway.

Perhaps the most welcome wire came to Wilkins from the *New York Times* with an offer to buy his exclusive story of the flight. He had spent most of his money paying the Barrow Eskimos for shoveling runways. Now they would have enough to buy food and lodging when they returned to civilization.

Spitzbergen was locked in ice. There was no aviation gas at Green Harbor, so they couldn't fly the Vega anywhere. They remained guests of the radio station until May when ice allowed the famed Norwegian ship *Hobby* (which had been used by Amundsen and other explorers) to stop near on its way to the Norwegian mainland.

There was enough gas left in the Vega for Ben to take off and immediately land on the ice beside the *Hobby*. The plane was hoisted aboard on May 10.

The two had little idea of the reception the world would give them as a result of their flight. They were greeted and celebrated in Norway, Denmark, Sweden, Germany, the Netherlands, Belgium, France, and England, and the United States.

When they arrived at Tromso on May 15, the docks were crowded with cheering people waving Norwegian, American, and Australian flags. The flyers were carried to the Grand Hotel on the shoulders of young men. There were banquets, crowds, and speeches. Ben used his Norwegian, which pleased the crowds.

A similar celebration in their honor took place at Trondheim. More crowds greeted them at Bergen, where they dined with King Haakon. The king presented Ben with the Leif Erickson Memorial Award, and the flyers were guests of the king at his country home.

At Oslo they were given another rousing welcome while six airplanes circled overhead dropping flowers. The Norwegian Aero Club presented them with medals.

Also at Oslo they were welcomed at the country home of Amundsen, where a huge cake had been prepared with a frosting map showing the route of the Vega in red. Amundsen presented each of the fliers with a Norwegian Aero Club Medal of Honor; he gave Ben a tie pin set with a stone he had collected on his South

Pole expedition. The old explorer told them that their flight was... "the most important ever made over unexplored land."

While they were guests of Amundson, word came that the dirigible *Italia*, returning to Spitsbergen from a flight over the North Pole, had crashed on the ice. A gondola had ripped off. The lightened ship had climbed back into the air and disappeared with six crew members still aboard. Explorer Umberto Nobile and part of the crew had been dumped on the ice, and were lost somewhere north of Kings Bay, Spitzbergen. These details came from Nobile, who managed to make a radio connection from where he was stranded. A huge international rescue operation was launched.

Amundson flew off to join the search for Nobile and his surviving crew. He never returned. The catastrophe cost, directly or indirectly, seventeen lives, including that of Amundsen. In the end, Nobile and seven of his companions were rescued.

At Copenhagen, Crown Prince Frederick feted Wilkins and Ben at a banquet. Both received gold medals from the Denmark Royal Aeronautical Society.

Crowds welcomed them at Berlin, Amsterdam, and London, where they collected more medals, gifts and memorabilia. At London, L.S. Amery, Secretary of State for the Colonies (British), said, "Never before in history has so much of the world's map been cleared up in one day."

George Hubert Wilkins was knighted by King George of England and became Sir Hubert Wilkins.

At Paris, Ben was called "An internationally famous arctic pilot."

Both men received high honors from the American Geographical Society, the Royal Geographical Society (England), and other scientific groups in various countries.

The welcome they received upon arrival at New York was similar; Stefansson was there, as was Dr. Isaiah Bowman of the National Geographic Society, as was a representative of President Coolidge. There were reporters. Always reporters.

In downtown New York they were showered with ticker tape as they rode in an open car. There were banquets, gifts, honors.

Later in the year Ben was awarded the Distinguished Flying Cross, which was presented at Washington, D.C. He was also presented with the First Flier of 1928 trophy by the International

League of Aviators, at Paris, France. His name is prominent in the National Aviation Hall of Fame list, at Dayton, Ohio.

Their Vega had been shipped back to the Unites States, and, with Wilkins, Ben flew himself and Wilkins on a wide-ranging tour of cities in the eastern and Midwest United States. Crowds greeted them. There were banquets, gifts and mementoes.

10 Flights in Antarctica

WHILE ON TOUR after returning to the United States from the Spitzbergen flight, Sir Hubert Wilkins announced that in the fall he was leaving for a three or four man expedition to the Antarctic with two Lockheed Vegas. He planned an aerial search for sites for twelve meteorological stations he hoped to establish. He was to live at Deception Island aboard the *Hectoria*, a Norwegian whaling ship, and would fly the Vegas from the sea ice there. He asked Ben Eielson to accompany him as first pilot.

On September 22, 1928, Sir Hubert Wilkins, with Eielson as First Pilot and three others—Joe Crosson—who had impressed Wilkins by his flight with reporter A.M. Smith from Fairbanks to Barrow in 1927—as Second Pilot, Orval Porter, mechanic, and William Gaston, radioman—left New York on the liner *Southern Cross*, bound for Montevideo, Uruguay. It was only six months and one week from the day Eielson and Wilkins had departed from Barrow for their Spitzbergen flight.

Also aboard was the Spitsbergen-flight Vega X3903, now named *Los Angeles*, plus a second Vega, named *San Francisco*. The second ship was equipped with a 450 hp Pratt & Whitney Wasp engine.

After the success of the Spitzbergen flight, Wilkins had no difficulty getting financing for this expedition. Included was newspaper publisher William Randolph Hearst, who paid $25,000 for exclusive reports on results of the venture.

The former White Star liner *Hectoria*, now a mother ship for six whale catcher ships, took the expedition aboard at Montevideo and

(L. to R.) Ben Eielson, First Pilot, Orval Porter, mechanic, and Joe Crosson, Second Pilot, the crew that accompanied explorer Sir Hubert Wilkins to Antarctica to make the first flights over that continent.

headed south. On arrival at Deception Island, where the *Hectoria* was based, they found insufficient ice in the bay there for a landing area for the planes, as Wilkins had planned.

The *Los Angeles*, mounted with pontoons, was launched, and Ben attempted a take off, but clouds of small birds interfered with the plane. At take off speed, bird strikes could damage the propeller, or even the windshield. He aborted the take off, and the plane was taken ashore and wheels were mounted.

A marginal runway was located in a snowy area. Skis were attached, and, with Wilkins aboard, on November 22, Ben took the *Los Angeles* off on a brief trial flight. It was the first airplane to fly in Antarctica.

Snow melted on the marginal runway, which was now bare cinders; unsuitable for wheels or skis.

A cold spell froze ice on the bay. The explorers marked an area that appeared suitable for a runway. Ben flew the ski-mounted *Los Angeles* off successfully, but on landing, the plane skidded onto thin ice, broke through, and the nose, with propeller and engine, plunged into the water. The wings kept it on top the ice.

Ben climbed out, swam to the edge of the broken ice, and managed to squirm his way to solid ice.

The whalers brought ropes and planks, and in a long day's

work, the airplane was retrieved. Some engine parts were replaced for fear of salt water damage, and the rest were dried.

Another attempt to fly a pontoon-equipped Vega from the water failed.

Finally, on November 16, 1928, Ben flew the heavily loaded *Los Angeles* from a hilly and dog-legged runway they had painfully cleared, and with Wilkins navigating and making observations, they headed south.

They flew over Bransfield Strait and Trinity Island with its 6,000-foot peaks. They reached Graham Land, where, eight years earlier, Wilkins had spent three months to map a forty-mile line; they now covered the same distance in twenty minutes. Peaks on Graham Land, they found, were more than 9,000 feet high.

When they were 600 miles from Deception Island, with half of their gas gone, they turned back. Wilkins estimated they had reached 71 degrees 20 minutes south, and 64 degrees 15 minutes west.

If they had been forced down, their chances for survival would have been low; no shelter or help was within hundreds of miles.

The Lockheed Vega San Francisco, *NC-32M, shown here, with the Spitzbergen flight Vega X3903, named* Los Angeles, *were the first airplanes to fly in Antarctica.*

They had enough food for a short time, but probably not enough to carry them 600 miles. If the second Vega was used in a search, the great distances, and the immensity of the land would have made it unlikely they would have been found.

They flew through a storm on their return. After nearly eleven hours of flight Ben landed safely at the dog-legged, hilly runway at Deception Island.

Stormy weather kept the planes grounded until January 10, when Wilkins, flown by Joe Crosson, made a 250-mile exploration flight searching for a more southerly landing area that could be used for a base in the next season of exploration that Wilkins was already considering.

That ended their season. The two Vegas were mothballed on Deception Island, the engines greased and the fuselages covered with canvas. The wings were removed and stored in a warehouse at the whalers' station.

Later, Wilkins donated Vega X3903, the airplane that flew from Barrow to Spitzbergen, and the first plane to fly over Antarctica, to the Argentine Republic. They agreed to place it in a museum. Instead, it rotted at Moron Airport in Buenos Aires.

In April, 1929, while he was in Washington, D. C., Ben was presented with the Distinguished Flying Cross, and President Hoover presented him with the Harmon Trophy for 1928.

Wilkins had plans for other explorations, and Eielson could have accompanied him. However his dream was to see aviation serving Alaska. He was convinced that Alaska would eventually become a major hub for air routes between America and Asia, and between Europe and Asia.

When he parted with Wilkins, he convinced the Aviation Corporation, which owned American Airways (later American Airlines) to buy three pioneer air services in Alaska (Anchorage Air Transport, Rodebaugh-Bennett, Wien Alaska Airways) and start a new company, Alaskan Airways Incorporated. He accomplished this in early 1929; Eielson was named vice president and general manager.

Wilkins' Career (condensed)

An inveterate explorer, in 1931 Wilkins captained a U.S. Navy World War I S boat (submarine) he named the *Nautilus* which was transferred to him without cost, provided he scuttle it after completion of his Arctic exploration. He navigated it under and on the surface of the Arctic Ocean to latitude 80 degrees 15 minutes north.

From 1932-39, he was involved in various Antarctic and Arctic expeditions, and from 1942-52, he was a consultant to the U.S. Army military planning division.

He was a geographer for the research and development command of the Department of Defense starting in 1953. In 1955, he received an honorary Doctor of Science degree from the University of Alaska Fairbanks.[1]

In 1947, during an expedition led by U.S. Navy Captain Finne Ronne, a group of mountains in Antarctica was named for Wilkins, and a prominent cape was named for Eielson. There is also Wilkins Sound, and Wilkins Ice Shelf, both in Antarctica.

Sir Hubert Wilkins died in 1958 at age 70 at Framington, Massachusetts. His ashes were scattered at the North Pole in March, 1959, from the surfaced U.S. Navy's atomic-powered submarine *Skate.*

---

1. [Author] I was present at that Commencement, hired by the University with my 4 x 5 Speed Graphic to take photos. I photographed Wilkins as he stood in cap and gown with faculty members and others next to the Eielson Memorial building. The prints and negatives (black and white) were retained by the University. JR

Book Three

The Siberian Challenge

11 Travails of the *Nanuk*; The Hamilton Disappears

AFTER LEAVING THE *Elisif* in October, 1928, and making his remarkable circumnavigation of the globe, Olaf Swenson reached Seattle in time to outfit the schooner *Nanuk* for its annual summer journey north. He needed to return to North Cape with supplies for the iced-in *Elisif*, and he needed to bring out the fur it still held. Furs the Russian government had collected at various Siberian coastal stations also awaited pickup.

THE 1929 VOYAGE OF THE *NANUK*

Accompanying Swenson on the 1929 voyage of the *Nanuk* was his pretty, bubbly 17-year-old daughter Marion, who signed on to the ship's company as a cadet. Her father's tales of the voyage, and the locking in ice of the *Elisif*, and his great overland trek, had excited her adventuresome spirit.

"I hope we do get frozen in," Marion replied when a reporter asked her if she was afraid the *Nanuk* would get stuck in the ice.

Robert (Bob) Gleason, a licensed radio operator, and a senior studying electrical engineering at the University of Washington, (for more on Gleason see Chapter 20) was to operate a newly-installed radio on the *Nanuk* (Morse code, no voice; call letters WKDB). His pay was $115 a month plus room and board (a bunk on the ship). Gleason and Charles Huntley, the radio operator on the *Elisif*, were long-time friends.

As the radio operator on the *Nanuk*, Gleason was the intelligence center of the ship. He sent and received every message to and from the *Nanuk,* whether personal from or to those aboard, or having to do with the fur or ship business. He also copied press reports that were broadcast by powerful transmitters in San Francisco and San Diego, which kept those aboard abreast of national and international news. He occasionally sent press releases having to do with the *Nanuk* and her status. He monitored radio transmissions from Alaska to the States.

The ship's owner, tall and personable Olaf Swenson, was fluent in Russian, and Gleason learned to copy Russian transmissions for him. He had frequent contact with the Russian steamship *Stavropol,* en route to the Arctic from Vladivostok, providing the ship with ice conditions near the *Nanuk.*

Gleason had aboard his personal five watt shortwave ham transmitter, with which he worked radio amateurs in San Diego and Honolulu.

The Captain of the *Nanuk* was R. H. Weeding, a small man of sober mien who had previously sailed with Swenson. The Chief Engineer was Bill Bissner, an Arctic veteran of about 60, also a long-time associate of Swenson. His Assistant Engineer, George Hunter, who had previously sailed in the Arctic, was about half Bissner's age.

The First Mate was Holmstrom, a bad-tempered Swede who had no arctic experience. Second Mate was Arnold Draven. Two of the four sailors, hired by Captain Weeding, had never worked on a sailing ship.

The Cook/Steward was Clark Crichton, whose 15-year-old son, Clark Jr., was the Cabin Boy.

Last was Tzaret Berdieff, an Americanized clerk of Russian descent who interpreted Russian, and who tallied deliveries of trade goods as they left the ship, and the furs as they were loaded.

The *Nanuk* left Seattle on the evening of June 15, 1929. Ten days later, under sail, she reached the Aleutians, transited Unimak Pass, and docked at Dutch Harbor.

On June 29, *Elisif* reported she was finally free of the ice that had imprisoned her for eleven long months. She anchored, and the *Nanuk* joined her shortly to transfer to her trade goods and supplies. After jockeying among open ice floes, to pick up furs and to trade, both the *Nanuk* and *Elisif* started along the Siberian coast from North Cape. The *Nanuk* soon passed the *Elisif*, for it drew less water, and was able to travel in ice-free water nearer the coast.

Loss of the *Elisif*

That August 11, 1929, the *Elisif,* damaged by ice, started to sink and was beached at Cape Billings, Siberia. She was a total loss. How the crew survived to reach 500-mile-distant American Diomede Island in small boats is one of the great sea stories of all time.

Captain Evan Larsen and his Norwegian crew of sixteen, plus Ray Polister, the Swenson Fur Trading company representative, Captain Jochimsen the arctic ice pilot, and radioman Charles Huntley, left the wrecked *Elisif* at Cape Billings on August 18 in two double-ender launches towing two thirty-foot-long barges. One launch had a fifteen hp engine, the other a twenty-five hp engine. They had enough gasoline for the 500-mile trip they planned.

Weather was good for five days and nights, during which they traveled day and night to reach the Chuckchi village at North Cape. After resting there, they pushed on and on August 27, reached the village of Uelen on the north side of East Cape, Siberia. There they remained, expecting to be picked up by the American Coast Guard ship *Northland.* However, on March 28 a fierce storm hit. Huge swells, with ice moving toward the beach, threatened their boats.

They quickly got the boats off the beach and offshore, and tried to enter the calmer lagoon there. However, they encountered high surf at the entrance bar. After nearly losing one barge with five men aboard, they decided to stay offshore until the weather improved.

As the storm strengthened they decided to seek shelter behind twenty-mile-distant Russian-owned Big Diomede Island, in Bering Strait. As the seas grew, tow lines to the barges snapped twice. Recovering their tows and again getting under way took expert seamanship.

Within five miles of Big Diomede the tow line from the larger launch became tangled with its propeller. The barge and launch became partially swamped. The smaller launch was shipping seas and its engine was wet and running poorly. Darkness was upon them, the wind increasing. The *Northland,* anchored at Port Clarence, offshore from Teller, at the time reported hurricane force winds (Force 11 on the Beaufort scale).

The two launches and their towed barges became separated. The small launch was worked alongside its barge and seven men transferred from the barge to the launch. The barge swamped as the last man leaped to safety.

After hours of searching, those in the small launch located the still afloat big launch and barge. Captain Larsen had rigged a sea anchor, and while the big launch faced tremendous seas, one of the crew plunged into the icy water to untangle the rope from the propeller.

The small launch was secured astern of the still-afloat barge where it rode out the night, while radioman Huntley and the chief engineer bailed to keep it afloat.

On August 30 as the storm weakened they found they had been blown sixty or seventy miles to the north. They abandoned the remaining barge, crowded all twenty of the men into the two launches, and headed for American Little Diomede Island. The sea was very rough, and both launches were overloaded, but they arrived safely off the village near dark. To end their great adventure, they had to run the launches through a huge surf to reach the rocky beach, where the Eskimo villagers waited to help.

The *Northland* arrived next day to pick the men up and take them to Nome, where they arrived September 2.

The *Nanuk*'s Voyage

The *Nanuk* continued to fight ice, and with some difficulty reached the Kolyma River and traveled upstream to the town of Nizhnekolymsk. There the crew unloaded trade goods and took aboard a fur cargo with an estimated value of $1 million.

With some delays due to ice, the *Nanuk* returned southeast, heading for Seattle. She reached the beached *Elisif*, where the crew salvaged fur, gasoline, kerosene, canned milk, coal, kerosene lamps, radio apparatus, as well as the dog team and sled the crew had used the previous winter while frozen in. Local Natives had cared for the dogs until the *Nanuk* arrived.

Nanuk Frozen in

On October 4, 1929, while bound for Seattle, the *Nanuk* became frozen in ice near North Cape, close to where the iced-in *Elisof* had spent eleven months. Ice was to hold the *Nanuk* there until July 8, 1930.

The trading schooner Nanuk, *locked in ice, near North Cape, Siberia, was loaded with valuable fur in November, 1929, and it appeared the price of furs might fall. Owner of the* Nanuk, *Olaf Swenson, agreed to pay the newly formed Alaskan Airways $50,000 to fly the fur and some of the crew of the* Nanuk *to Fairbanks. The fur could then be transshipped via rail and sea to New York fur markets. Ben Eielson, General Manager of the new company, estimated six flights would do the job.*

Olaf Swenson, an old hand in the Arctic, was prepared. The *Nanuk* was converted to a home on the ice. A gangway was built so all could easily leave the ship and walk on the surrounding ice to shore, or to make a two-mile walk to visit the *Stavropol*, a Russian ship which had arrived nearby, and was now also a recent prisoner of the ice.

A tarpaulin was stretched across the foremast boom of the *Nanuk* and tied to the bow and side rails, giving a clear and sheltered work space on deck. A shack with a tight self-closing door was built on deck over the main companionway that led below. An outhouse was built projecting over the ice. All hull openings were blocked, and pipes were drained so they wouldn't freeze and burst.

Chief Engineer Bissner built from five-gallon metal cans three stoves designed to burn diesel, of which there was an abundant

A Stearman (left), and two Fairchild 71s are parked near the Alaskan Airways hangar at Fairbanks/ Weeks Field, circa 1929. The hangar had just been purchased from Wien brothers as the new company was organized. Among the airplanes owned by the new company was the former Wien Hamilton Metalplane NC10002, as well as the Stinson Detroiter biplane the Wien brothers had owned.

supply aboard. A fifty gallon drum was cleaned and kept in the galley for fresh water. Ice, cut from the surface surrounding the ship, was carried aboard and placed in the drum (old sea ice, when exposed to air produces fresh water).

The fur market was high at the time, but there were indications it wouldn't last. Swenson decided to make every effort to get the fur off of the ship and on its way to the New York and London markets. He also wanted to get some of the *Nanuk's* crew away; a small ship wintering in ice can lead to crew conflicts; the fewer aboard the better.

The flight by Noel Wien to the *Elisif* the previous March had shown that air service from Alaska was possible. With the ship's radio, Swenson offered Alaskan Airways, $50,000 to fly the crew of the *Nanuk* and $1 million worth of fur to Fairbanks from where it could be shipped by rail and sea to the New York and London fur markets.

The New Company

The new Alaskan Airways Incorporated company owned ten airplanes, plus two on the way from the factory. Ben Eielson was general manager. The pilots were Ed Young, Noel Wien, A. A. Bennett, Ralph Wien, Russ Merrill, Frank Dorbandt, and Tom Girard. Girard was supervisor of maintenance as well as a pilot.

Noel Wien Leaves Alaska

Noel Wien, no longer owner of a flight service, could have remained as a pilot for the new Alaskan Airways, Inc. However, he and Ada, his wife of five months, instead took a trip to the states. They had never taken time for a honeymoon, and Noel wanted to check out flying jobs in the South-48. At this point he had logged 2,440 hours of flight time, nearly 1,800 of it in Alaska.

The aviation world in the United States had changed since Noel's arrival in Alaska in 1924. He couldn't have picked a better time to look for a job as a pilot.

Sadly, while Noel and Ada were in Minnesota, Noel's brother Ralph was killed at Kotzebue, Alaska, as he flew a Christian mission-owned Packard diesel-engine-powered Bellanca airplane. The plane banked about thirty degrees to make a landing approach when, suddenly, from about 400 feet it plunged into the ground. Father W. F. Walsh, head of the Kotzebue mission, and Father Philip Delon, his passengers, were also killed.

There was no clear cause of the crash. Noel thought a passenger became frightened when Ralph banked, grabbed the duel controls, and froze. Sam O. White, a flying game warden, and a close friend of the Wiens, thought that something rolled into the controls to jam them when it banked.

The Wien family was close. The loss of Ralph was keenly felt by the family, and by his many Alaskan friends.

Commercial Aviation Comes Alive in U.S.

The age of commercial aviation in the United States was born on May 20, 1927, when, that morning, a young man, called Slim by his friends, urged a tiny Ryan monoplane into the air from a muddy New York runway. The engine roared, men pushed on the struts, and the wheels left deep tracks. Slowly, the overloaded plane gathered speed. Then the pushers couldn't keep up. The cutoff

Newly married Noel and Ada Wien, with a Standard biplane that had an enclosed front cockpit. When the Wien brothers sold their company to Alaskan Airways, Noel and Ada traveled to the states on a delayed honeymoon. They later returned to make Alaska their permanent home.

point passed—Slim was now committed; there wasn't enough runway left for him to stop even if he tried. He had to fly, or crash when he ran out of runway. If he flew, he had wires to clear.

He held the plane down, allowing it to gain speed. Then, carefully, he lifted it. He cleared the wires by twenty feet, and was on his way to Paris in what was to become the most famous airplane flight ever.

Charles "Slim" Lindbergh was probably the most celebrated hero in modern world history. No one questioned the right of it when President Calvin Coolidge dispatched the Navy cruiser *U.S.S. Memphis* to bring him home from Europe. More than 500 vessels welcomed him when the *Memphis*, with him aboard, arrived in New York Harbor. To list medals and other honors he received would require pages of small type. From sixty-nine countries, he received 15,000 gifts worth two million dollars. Three million congratulatory letters, and a quarter of a million cablegrams arrived and were loaded on a truck that followed him in a ticker-tape parade through New York City.

Soon, thanks to Charles Lindbergh, no longer did most of the flying in the United States consist of colorful barnstormers, or flying circuses, with pilots barely eking a living with WWI surplus Jennies, Standards, and the like.

Overnight, boys and young men decided to become pilots. New airplane factories sprang up in the United States, and modern cabin planes became common. The Travel Air Company, at Wichita, Kansas, started in 1925 by Clyde Cessna, Walter Beech, and Lloyd Stearman, all three still famous names in aviation, led the way. In 1929, one-fourth of all registered airplanes in the U.S. had been built by Travel Air. We as a nation had started to catch up to aviation-minded Europe.

In the 1930s, many pilots seeking to establish records flew across oceans, from country to country, and between major cities. Speed and altitude records were set. Pilots who participated became famous. The names of American pilots Frank Hawks, Jimmie Doolittle, Charles "Speed" Holman, Roscoe Turner, Bernt Balchen, Wiley Post, and others were constantly in the news. Prize money and trophies like the Bendix Cup (1931–49), the Thompson Trophy (1930–49), the Schneider Trophy (1913–31), the James Gordon Bennett Cup, and later, the National Air Races, drew huge crowds and their results filled pages of newsprint.

By his grand flight, Lindbergh demonstrated to the world the potential of air travel and transport. Passenger, freight, and airmail service increased, and airplanes were more and more being used for purposes other than entertaining crowds that both hoped and feared they would see some daredevil pilot killed.

Today Lindbergh's *Spirit of St. Louis* hangs above the entry to the Smithsonian's Air and Space Museum in Washington, D. C., as one of the two most significant historical airplanes it owns (the original Wright Flyer is the other).

Late in his life, dying from cancer (he died in 1974), Charles Lindbergh slipped into the Air and Space Museum after closing time one evening to find, as he had requested, the *Spirit of St. Louis* had been lowered to the floor. Museum personnel gave him privacy while he walked around the old airplane as if giving it a preflight. He climbed aboard and sat in it for an hour. One can only imagine his thoughts.

One-time barnstormer and long-time Alaska bush pilot Rudy Billberg told me (author) that after Lindbergh's flight, the Ryan factory received many orders for copies of the their NYP model (New York to Paris), minus the extra fuel tanks, and with visibility through a windshield (except for a periscope, the *Spirit of St. Louis* had no forward visibility).

Billberg flew one of these NYP models. "Moving the control stick in flight took a lot of strength", he told me. "One was tempted to add a length of pipe to it to gain leverage. Also, the airplane was so unstable in flight that a pilot had to continually correct."

Perhaps it was designed that way to help keep Lindbergh awake for those long hours over the Atlantic Ocean. That airplane was designed for a single purpose – to fly the Atlantic.

Any recounting of 1920s and 1930s aviation in the United States should include a salute to Charles Lindbergh, for he lit the fire that resulted in the great aviation jump forward made by the U.S., including Alaska.

In 1929, when Noel Wien temporarily left Alaska, thanks at least partly to Lindbergh's flight, good steady jobs were available in the South 48 for experienced pilots.

Before leaving Alaska, Noel, who had flown the Hamilton NC10002 for 425 hours, checked Ben Eielson out in it. During two check rides, Ben had difficulty landing the plane, which worried Noel. Eielson didn't want Noel to leave Alaska. He thought no one could fly the Hamilton as well as Noel.

Flights to the *Nanuk*

Alaskan Airways and Swenson's agreement was signed, and the Soviets issued a permit for Alaskan Airways to fly to the *Nanuk* to remove fur and passengers. On October 26, 1929, Bob Gleason, the *Nanuk* radioman, received a message:

> Frank Dorbandt is at Teller, waiting for weather to clear before flying to the *Nanuk*. Ben Eielson will follow from Nome with the Hamilton.

Flying from Fairbanks, Eielson damaged a landing gear of the Hamilton while landing at Nome. He was delayed while it was repaired.

October 29 was a beautiful clear day both at Teller and North Cape. Gleason, on the *Nanuk,* received a message:

> Dorbandt has left Teller flying the Stinson Detroiter, bound for the *Nanuk.*

The airplane was the *Stinson Detroiter Number 2*, the four-place biplane the Wiens bought from explorer Hubert Wilkins and which for two years they had flown year-around between Nome and Fairbanks.

The Siberian North Cape Eskimo village of about 100 persons included local Natives, two Russian families, two wood frame houses, and many Native driftwood-sod huts. Word of the expected airplane soon spread. The *Nanuk's* crew and all of the villagers stood on the ice looking and listening for the first sight and sound of Dorbandt's airplane.

Six hours and twenty minutes after leaving Teller, six-foot, powerful, Frank Dorbandt, accompanied by mechanic Bud Bassett, neatly landed the ski-equipped Stinson on the ice, and taxied to the *Nanuk.* The crew loudly cheered his arrival. That airplane and its pilot was their link to home, to civilization, to another world. The prospect of living on the ice-bound ship through the winter had not been well received by all. Perhaps Marion Swenson looked forward to it; she was a cheerful soul and brightened the atmosphere of the ship.

Dorbandt's load was mostly gasoline for his return flight, although he did bring mail, cigarettes, and a case of coffee—all light items.

Dorbandt, who had a very high regard for Dorbandt, immediately filed a message with radioman Gleason, who, at Dorbandt's insistence, sent it to the Associated Press in New York:

> Pioneer Alaska pilot Frank Dorbandt today completed a hazardous six-hour-twenty-minute flight from Alaska to North Cape, Siberia, to rescue the crew of the *Nanuk,* which is frozen in the ice there.

Dorbandt waited for Eielson to arrive, so the two planes could return to Alaska together.

Eielson, with mechanic Earl Borland, arrived at the *Nanuk* the next day after flying the Hamilton Metalplane four hours and twenty minutes from Nome. The ice in the bay was smooth. Earlier, Dorbandt had Gleason radio a message to Eielson to leave wheels on the big plane instead of changing to skis. Eielson made a nice landing. There was little snow, and a wheel plane could safely land almost anywhere in the bay.

The *Nanuk's* crew was immensely cheered by the presence of Eielson and Borland. Bob Gleason later wrote, "We had a great evening with these men. Eielson and Borland, contrasting with Dorbandt's loudness, were both quiet spoken, calm, and most pleasant."

Later, Gleason commented on Dorbandt, "We had come to feel he was mainly a publicity hound."

Planes Return to Alaska

Both planes left, bound for Alaska, the next day, November 1, taking with them six members of the fourteen-member *Nanuk's* crew. Four rode in the Stinson—the Assistant Engineer, Cook/Steward Clark Crichton and his son Clark Chrichton Jr.,[1] and one of the sailors. Eielson's plane carried one sailor and the unpleasant First Mate Holmstrom. The Hamilton was also crammed full of fur, and its 525 hp Hornet engine swiftly yanked it into the air despite the big load.

The two planes ran into heavy falling snow near Siberia's Cape

1. Years later, Clark Crichton Jr. joined Gleason in Alaska in the airline communications business.

Serdze and were forced to land on smooth ice near a Chukchi village. While the blizzard raged, the hospitable Natives housed the visitors for three days. The two planes left on the fourth day; Eielson landed at Nome, Dorbandt at Teller.

Second Flights

Waiting for return flights to the *Nanuk* for more fur and passengers, Eielson and Dorbandt, with their mechanics, had their planes fueled and ready. They were at the Lomen roadhouse in the tiny coastal Seward Peninsula village of Teller, stymied by persistent blizzards and fog.

Jack Warren,[2] a sturdy, friendly and capable sourdough, with his wife Helen, managed the Roadhouse. A big man with the nose of a pug, Warren had been a professional boxer before coming north. He became a licensed airplane and engine mechanic, and later worked in Fairbanks for Alaskan Airways. He had once been in charge of the thousands of Lomen Brothers' reindeer at Nome, before the federal government bought all the non-Native-owned deer.

He is one of the unsung heroes of the aerial drama that was about to take place. Pilots and mechanics depended on Jack and Helen Warren and the roadhouse for living necessities and operational support. Impatient pilots and frustrated mechanics, irritable under the strain of terrible weather and the need to fly, were pacified and encouraged by Jack, who didn't reveal his own stress, something in which he took great pride afterward.

That winter of 1929–30 was one of the stormiest Jack could remember. "Nothing but snow, fog, low temperatures, and wind," he said.

On November 9, after waiting five days, Dorbandt, a braggart and a loudmouth, who commonly mercilessly heckled others, was restless. He was a capable pilot, but he was impulsive, sometimes reckless. One of his quirks annoyed other pilots and frightened passengers. Upon arriving at a destination he sometimes flew a violent maneuver—a wingover, maybe a loop. It was his way of saying, "Here's the great Dorbandt."

2. The author was privileged to meet Warren in 1953 when he lived near Big Delta. At the time I had no idea of the part he had played in this 1929-30 aviation drama. He was a friendly, jovial, true sourdough. Jack Warren Road, near Delta Junction, is named for him. JR.

That day he demanded of Eielson, "When are we going to go?"

Eielson, his boss, was calm and patient. "We need another weather report from the *Nanuk,*" he said, quietly.

"There's nothing to it," Dorbandt bragged. "I don't know why we're waiting."

Eielson didn't answer.

"You can sit here if you want to. I'm going," Dorbandt declared, defiantly.

He dressed in his flying clothes, and with Bassett, headed for the Stinson Detroiter.

Fog lay over Bering Strait. Snow was falling. A storm was in full swing. At 10:15 that morning those in the roadhouse eyed each other uneasily when they heard the roar of the *Detroiter*'s Wright Whirlwind as Dorbandt flew it into the storm.

Eielson quietly got into his flying clothes, and about an hour later, with Borland aboard, flew the Hamilton into the storm. Speculation continues to this day about his decision to fly. Was Dorbandt's goading responsible for Eielson's decision to fly? Dorbandt reportedly once called him a coward, which must have rankled. Or, was Eielson worried that Dorbandt might get into trouble? Two planes together were safer than one; if one goes down, at least those in the other plane might know where it landed. A downed plane on the drifting ice of Bering Straits in November was about as certain a way to death as one could find.

Dorbandt soon flew the Detroiter back to Teller. Despite his brags, the heavy snow and fog were more than he could handle. "No pilot could cross the Bering in this weather," he gruffly explained.

The Hamilton Disappears

The Hamilton didn't return, nor did it arrive at the *Nanuk.* The big plane carried food and emergency gear. In that year of 1929 it wasn't uncommon for an Alaskan pilot to be forced down due to bad weather. There were no weather forecasts in the Territory, and it's easy to fly into unreported bad weather. Airplanes had no radios.

"Pilots walk more miles than they fly," claimed some who doubted the future of flying in Alaska. It was true that many an Alaskan pilot, forced down in the wilderness, had to walk to the nearest village or town. Many thought it likely that Eielson and

The former Wien Hamilton Metalplane NC 10002. Note the "I", part of the former Wien company paint job that still appears above the Alaskan Airways painted on the fuselage. This is the airplane in which pilot Ben Eielson, with his mechanic Earl Borland, flew into a Bering Sea storm to never return.

Borland would walk to a coastal Native village from their downed plane.

At first there was little concern. But, as time passed without word of the Hamilton or its crew, alarm grew. Talk of a search began. It was a terrible time of year for an aerial search. In November on Alaska's Seward Peninsula, and across the Bering Straits in Siberia, there are only three to five hours of enough daylight for safe flying. The sun disappears for the winter at North Cape on November 15. By December, daylight hours are even fewer. And, as Jack Warren commented, the weather was terrible. In fact, it was to become one of the stormiest winters on record for the Siberian coast, the Bering Straits, and northwest Alaska.

Daily reports from the *Nanuk* repeated the grim message:

> No sign of the Hamilton. Heavy snow. Visibility nil.

At first reports of the missing plane appeared in the *Fairbanks News-Miner* on November 12 under "Aviation Notes," which listed the coming and going of various bush planes and their passengers at Fairbanks:

> No word has been received from the Alaska-Siberia transport planes piloted by Ben Eielson and Frank Dorbandt since the news arrived that they were ready to hop from Teller to Cape North on the second trip.

By November 13, the *News Miner* reported that two Anchorage planes that were supposed to fly to Siberia to search for the Hamilton had been damaged in take-offs, and wouldn't get to Siberia. Dorbandt had damaged the left axle of the Stinson Detroiter in attempting a takeoff on hard-packed, rough, snow. He repaired the damage himself.

By November 30, the search for Eielson was front page news in the *News-Miner*, with big print:

> STRONG WINDS WITH LOW VISIBILITY PREVAIL AT NORTHERN SIBERIA POST. Land Searchers Fail Find Lost Fliers. Crosson and Dorbandt Will Go By Air With First Break in Weather. Eielson Flying Off Shore When Last Seen by Trapper.

The *News-Miner* reported on a wire from Teller;

> Dorbandt said that Olaf Swenson, aboard the *Nanuk*, had wired that just as it was getting dark with bad visibility, Eielson circled around a Native's house twice about sixty miles from here [the *Nanuk]* but did not land.
>
> A Russian trapper heard the plane about fifty miles from here [the *Nanuk*]. Cape Serge Natives who have just arrived report there has been no visibility for the last few days. Think Eielson is somewhere along the lagoon east of here. Yesterday the visibility was quite good. Possibly Eielson damaged his landing gear. Otherwise should have arrived. If weather permits tomorrow will send dog team in an endeavor to locate him.

Other newspapers across the United States, as well in Europe, carried the story of the missing Eielson and Borland. Over the previous two years, Carl Ben Eielson's name, detailing his flights seeking land in the Arctic, including the flight to Spitzbergen, and his flights in Antarctica, had often made international headlines. Now his name was again in the headlines, and it would appear there sporadically for more than two months, until he, Borland, and the Hamilton were found.

12 The Search for the Missing Fliers

THROUGHOUT THE WINTER of 1929–30, storm after storm swept across Siberia, Bering Strait, and northwestern Alaska. Winds roared, heavy snow fell, impenetrable fog lay upon the sea, and the temperature often plunged to -40 F. and colder. No pilot could fly in such weather.

Frank Dorbandt, waiting at Teller for flying weather, finally got it on November 16. He started to take off with the *Stinson Detroiter* C5262 heavily loaded with gasoline. The weight was too much for the plane. During the take off run on the rough, icy, landing strip, the biplane's axle cracked and a wing tip was damaged. He aborted the takeoff.

Dorbandt's mechanic, Bassett, tried to make repairs, but there was no machine shop; other mechanical facilities at Teller were also inadequate. A few days later the two men flew the lightened plane to Nome, where the axle was welded and the wing tip was repaired. They returned to Teller on the 26th, hoping for weather that would allow a flight to Siberia.

The storms continued. Day after day fog, snow, high wind, and far below zero temperatures prevailed. Dorbandt paced, peered constantly toward stormy Bering Strait, and kept the biplane ready.

Joe Crosson, flying open cockpit Waco 10 780E, arrived at Teller November 29, prepared to fly in search of the missing aircraft. Three weeks had passed since it had flown into the storm. In that time, Bob Gleason, the *Nanuk's* radioman, had reported only two flyable days at North Cape.

Natives who lived sixty miles from the *Nanuk* reported that the Hamilton had circled one of their houses twice on the day it had left Teller, but it didn't land. A Russian trapper heard it fly over on the same day but he didn't see it. He lived about fifty miles from North Cape. He thought it had headed inland.

To check out these reports, Olaf Swenson sent Tzaret Berdieff to search with the dog team they had recovered from the lost *Elisif.* He returned without finding anything of interest. At Swenson's request, Russians with dog teams also searched. Smoke seen by Natives along the coast where no Natives lived raised hopes that it was from a camp of the missing fliers. A ground search there found nothing.

On November 30, Crosson, flying the Waco, and Dorbandt with the Stinson Detroiter, took off and flew as far as Alaska's Cape Prince of Wales before heavy snow and high wind forced them to return to Teller. Weather at the *Nanuk* was -35 F., with a forty-mile-an-hour wind and no visibility.

Again, on December 6, Crosson and Dorbandt attempt to take off with both planes heavily loaded. Sastrugi (hard-packed wavelets of wind-driven snow) caused damage to both their landing gears and they had to abort. The *Detroiter* received most of the damage; Crosson's Waco was soon repaired.

While Crosson and Dorbandt waited at Teller for flyable weather, the Aviation Corporation (AVCO) owner of Alaskan Airways, arranged for three of the new and tough Fairchild 71 monoplanes to be shipped to Fairbanks from Canada and the states. They were to be assembled at Fairbanks and flown to Siberia to search for the lost fliers.

Once they were found, the new planes could complete the contract with the Swenson Fur & Trading Company by hauling the fur on the *Nanuk* to Fairbanks. A new design, they were powered by a Pratt & Whitney 420 hp Wasp engine, and had wings that folded for hangar storage. This was to be the first appearance of these seven-place (including pilot) transport ships in Alaska. They soon proved to be well suited for the Territory, whether flown on skis, floats, or wheels, or hauling freight or passengers, or both. Cruise speed was about 100 mph, with a 650-mile range. Payload with full tanks was 1200 pounds. In all, thirty of these rugged airplanes eventually flew in Alaska; the last ones flew into the 1950s.

Canadian pilot T. M. (Pat) Reid, a volunteer for the search, had just completed a six-month three-plane expedition of the Hudson

Joe Crosson, with his usual broad smile, with the Super Swallow biplane he often flew. He wears the furs that allowed him to fly in an open cockpit plane in winter.

Harold Gillam with the Wien Stearman biplane NC5415 he flew in Siberia while searching for the lost Eielson and Borland. Earl Borland was the mechanic who repaired this ship at Walker Lake after it was purchased by the Wiens in May, 1929. This airplane is now on display at the Alaska Aviation Heritage Museum in Anchorage.

Bay district and arctic Canada's Coronation Gulf. Canadian pilot G. Swartzman, who also arrived at Fairbanks with the Fairchilds, was also experienced in the Arctic.

Flying Alaskan Airway planes to participate in the search, Ed Young (See Chapter 17 for a review of Young's flying career) and Harold Gillam (See Chapter 16 for a review of Gillam's flying career) arrived at Teller December 11. Young flew as a pilot in France during World War I, and had been flying in Alaska for five years; Gillam, had worked as an aircraft mechanic for about a year, and had made only one cross country flight. He had a mere forty or fifty hours (sources vary) of solo time aloft. At Fairbanks he had begged, "Give me a ship. I want to search for Ben." When it seemed they were going to turn him down, he made his case by pleading, "If any pilot was lost out there, Ben would be the first to start looking for him."

He got his plane, the open-cockpit Wien Stearman biplane NC5415 (often referred to as "the Wien Stearman") which Earl Borland had spent two weeks repairing at Walker Lake the previous spring. Young flew a Stinson Standard (biplane) cabin plane.

On December 16, Alaskan Airways, Eielson's company, put prominent businessman Alfred J. Lomen in charge of the search for the Hamilton. Lomen, a Nome merchant and owner of valuable reindeer herds, had long outfitted miners and others who had challenged Alaska's far places. Frank Dorbandt had made the decisions until then, but he didn't work well with others, and Alaskan Airways managers had heard about it. At the same time, personable and well-liked Joe Crosson was named Chief Pilot of the operation; he had logged about 1,100 hours of flight time at this point.

On December 19, Dorbandt flew the now beat-up *Stinson Detroiter* C5262 to Nome. Dorbandt had flown it round trip to the *Nanuk* with Eielson and the Hamilton. Since then it had repeatedly needed repairs after taking off and landing with heavy loads on a rough, icy airstrip.

Harvey Barnhill flew it from Nome to Fairbanks where it was hangared and work started to bring it up to standards.

On December 22, while Alaskan Airways mechanic Ed Moore was cleaning the *Detroiter*'s engine with gasoline, a spark set the gasoline and the plane afire. As the fire spread, three other planes in the hangar were yanked to safety, but the old Stinson and the hangar were both lost.

It was a sad loss for an historic airplane. It was the fifth Stinson ever built, and it had made notable aviation history in Alaska when Wilkins used it on his third flight into the Arctic. With owners Noel and Ralph Wien it had been the first to fly commercially year-around, and make scheduled flights between two Alaskan cities; it was the second airplane to fly a round trip from North America to Asia.

Finally, on December 18, it appeared that flyable weather had arrived. Bob Gleason on the *Nanuk* reported good weather at the *Nanuk*, and it was good at Teller. Young, Barnhill (who had returned to Teller), Crosson, and Gillam all started for North Cape. Barnhill and Gillam returned shortly after running into heavy snow over Bering Strait; Crosson and Young followed them to Teller after flying for two hours.

Bad weather had again triumphed.

At Teller, on cloudless days there were now only four hours of daylight suitable for flying; less than that when it was overcast. The *Nanuk* was north of Teller, where there were even fewer hours of daylight. Heating airplane engines, repairs on planes, and everyday living took place mostly during the twenty hours of darkness.

The constant storms and the short daylight hours weighed on the searchers. It was now six weeks since the big Hamilton had disappeared into a storm. Since the earliest years of aviation in Alaska, when a pilot went missing, it was customary for other Alaskan pilots, regardless of company affiliation, to join the search. Searches always continued as long as there was hope the missing could be alive. Eielson and Borland were experienced Alaskans who were capable of surviving if uninjured, for the plane had emergency gear aboard, as well as cases of eggs, bacon, and ham ordered for the *Nanuk.* Food for them would not be a problem. The search would continue.

Joe Crosson decided when planes could reach Siberia the search would be made by the Waco 10 780E he had flown to Teller, and the Wien Stearman NC5415 Harold Gillam had flown. Both were open cockpit biplanes powered by Wright Whirlwinds. The Waco could carry two persons in the front cockpit, and with a top airspeed of 136 mph, was the fastest plane owned by Alaskan Airways. Both were faster than the larger cabin planes, and easier on gasoline, which would have to be flown to the *Nanuk* for any flying to be done from there.

Finally, on December 19, Joe Crosson and Harold Gillam left Teller with full fuel tanks and, flying in the twilight, crossed Bering Strait to Siberia, bound for the Chukchi village near Cape Serdtse-Kamen where Eielson and Dorbandt had spent two nights in early November. Crosson flew the Waco 10 780E; Gillam flew the Wien Stearman NC5415. From there it was still about 300 miles to the *Nanuk*. Each plane, in addition to full tanks, carried thirteen five-gallon cans of gas, a sleeping bag, a primus stove, and food for a month.

They took off from Teller in clear weather and flew at about 6,000 feet. Soon haze limited their visibility. They crossed Bering Strait, and flew past East Cape, Siberia. When they were opposite Cape Uelen they had to fly within 200 feet of the ice because of visibility-limiting fog. Soon the fog forced the two biplanes down to 100 feet. The two pilots peered down at the rough ice; all they could see ahead and on each side was impenetrable fog. Although they flew close together, they were out of sight of each other most of the time. They came to a dark bluff at the edge of the ice, and were blessed to be able to see and follow it. The end of the four hours of semi-daylight flying time was near, and they looked for a village where they could land and have shelter for the night.

They missed the village they had set for a goal, but tiny villages were found about every ten miles along the coast, and they landed at one called Pelikii, and felt fortunate to have arrived there safely.

The hospitable Chuckchi villagers helped them drain oil from the planes and secure them for the night.

That evening, through sketches and motions, the fliers learned from the villagers that they were near the southern end of Kolyuchin Bay. This put them about half way to the *Nanuk*. The villagers knew about the lost plane, and conveyed to the pilots that they had seen it fly over, heading toward the *Nanuk.*

Now Crosson and Gillam knew for certain the Hamilton had reached Siberia. That narrowed the search area.

Next morning, working in the dark, they fire-potted and warmed their engines. Thermometers lashed to their interplane struts read -36 F. There was a twenty-mile-an-hour north wind. Faint light arrived at 10 o'clock from a streak of light on the eastern horizon. It was December 21, 1929 (Siberian time—one day ahead of Alaska), the shortest day of the year, with barely three to four hours of light suitable for flying. The sun never rose above the horizon that day.

The two planes were aloft by 10 o'clock. To clear nearby hills, they climbed to 2,500 feet in light fog. Beyond the hills they descended through the fog in the dim light, searching for the shoreline, planning to follow it to the *Nanuk*. They then became separated.

Gillam circled, searching for Crosson. Crosson encountered dense fog and chose to return to the Chukchi village where they had spent the night. Fog was so heavy he couldn't see the ground for a time. He saw a pressure ridge, and flew just above it back to the village. As he circled to land, his goggles became fogged. He pushed them onto his forehead, and in the short time it took him to land his eyes became frosted. He again spent a night with the hospitable Pelikii villagers.

Gillam, peering in the direction of the village, decided the fog was too dense to return there. He was offshore by then and found an open lead, a dark streak of open water surrounded by ice. He descended to 200 feet, used the lead as a reference, and flew through fog, flying low to keep ground contact. Eventually he flew part of the time on instruments, following the shoreline to North Cape. His only map was a crude hand-drawn one given him by Frank Dorbandt. He recognized twenty-miles-apart Cape Vankarem and Cape Onmen and other features. Visibility was so poor that he flew into one low hill; it's slope was gentle, and he hit it at the right angle with his skis, so he bounced back into the air with no damage, but it gave him a terrible scare.

He landed at the *Nanuk* at 1:45 after an amazing mostly just-above-the ice three-hour flight in and out of shifting fog.

That flight by Gillam became one of the almost-legendary exploits of his spectacular and mostly charmed flying career.

Next day, December 22, Siberian time, an anxious Crosson spent three hours aloft bound for the *Nanuk,* searching the ice and shoreline for both the Hamilton and Gillam's Stearman. He feared the relatively inexperienced Gillam had become lost, or had been forced down. He was both amazed and relieved to see the Stearman parked serenely next to the *Nanuk.*

The greenhorn pilot had outflown him.

While Crosson rested on the *Nanuk* and cared for his frosted eyes, Gillam flew Olaf Swenson to foothills southwest of the ship where smoke had been reported a month earlier. They found a camp of reindeer herders, nothing else.

On the following day, Crosson, with Olaf Swenson in his front cockpit, and Gillam, accompanied by Demetri Miroshnishenko, a Russian government agent from the North Cape station, searched east of North Cape for more than two hours, and found nothing.

The pilots and the eight-person remaining crew of the frozen-in *Nanuk* celebrated Christmas with reindeer roast; Swenson had purchased several carcasses from inland reindeer herders for winter use.

Crosson and Gillam had local villagers build with blocks of snow a temporary shelter, essentially a nose hangar. It gave some protection to the engines of their planes. It helped, especially when they warmed the engines with firepots and the wind blew, which was most of the time. The shelters didn't keep windblown snow from drifting into all the cracks and crevices of the two planes, despite tight canvas wrapped around engines and cockpits.

In extreme cold it was impossible to fully warm the engines on the ground. The radial engines had no cowling around them – the cylinders were bare and exposed. With heated oil poured in, and the engines heated as much as possible with the plumbers' fire pots, they could start them. But, even while running, these engines soon cooled and stopped. To solve this, they took off immediately. At full throttle or cruise, they remained warm enough to continue to run.

Storms kept the two pilots on the ground until December 31. On that day Crosson flew another search for an hour and a half in the region east of North Cape. The engine of Gillam's Stearman died on takeoff, and he had to land dead stick on rough ice. Both landing gear struts were broken, the lower left wing tip was damaged, and the cabanes (struts between wings of a biplane) were damaged. Neither Gillam nor his passenger/observer Miroshnishenko were injured.

Alaska's acting governor, Carl Thiele, wired Alfred Lomen that, "We are authorized to ask for help from the Russian ice breaker *Litke*, and the Russian steamship *Stavropol*, in the search for the Hamilton." Secretary of Interior Wilbur had received an okay on this from the Soviet government.

The *Stavropol* was frozen in a short way from the Nanuk, and had already been helping. On December 15, the *Litke* was reported to be at Petropavlovak, at the tip of the Kamchitka Peninsula, roughly 1,500 miles from North Cape. With it were Junkers planes, which were expected to fly to North Cape as soon as the sun returned on January 18.

The Russians assigned locals with dog teams to the search, both from the *Stavropol* and from Wrangel Island, which lies 300 miles northwest of North Cape. They offered a reward of 2,000 rubles (about $1,000) for anyone with solid news of Eielson's location. Notice of the reward was sent to Siberian villages and reindeer camps.

This was an unusual concession from the Soviets, for Russia was not then recognized by the United States. It is probable that the respect and admiration of the Soviets for Eielson's arctic flying exploits had much to do with their cooperation.

The U.S. Army and Navy, when asked for help in the search, sent their regrets. "We have no pilots with experience in the Arctic." Ben Eielson was the only Army pilot with Arctic experience.

Far to the south, the U.S. Coast Guard, with the cutter *Chelan*, hustled two of the three crated Fairchild 71 airplanes AVCO had ordered, plus the Canadian volunteers, from Seattle to Seward in four days, arriving there on Christmas day. Planes and the Canadian crews were instantly loaded on Alaska Railroad cars at Seward, to arrive at Fairbanks the next day. Horse-drawn sleds waited near the rails at Fairbanks, and the crated planes were transferred to them within fifteen minutes of their arrival. They were assembled in the Alaskan Airways Inc. hangar (formerly Wien's). Major H.C. Deckard, of Fairchild Airplane Manufacturing Company, supervised the work, which continued around the clock.

The first Fairchild 71, fitted with skis, was flown on a test flight December 28 by Pat Reid and Forestry Patrol pilot G. Swartzman. The second Fairchild 71 was soon ready, and on January 2 they were both loaded and ready to fly to Nulato on the first leg to Nome and Teller.

Fairchild 71, Canadian registry CF-AJK, flown by Pat Reid, took off and was circling the field, waiting for the second Fairchild, U.S. registry NC190H, flown by Swartzman.

Anxious to get airborne, Swartzman hurried his takeoff and attempted it with a cold engine. The heavily-loaded plane became airborne, but couldn't sustain flight. It veered, the landing gear plowed through brush, and the plane ended on the ground about fifty feet from the end of the runway.

It had swapped ends. The left wing pointed at the sky. The right wing was crushed. The plane was a total loss; only the engine and a few instruments and other parts could be salvaged.

Fortunately, the third Fairchild had arrived and was ready for use. A new departure date was set for the next day, January 4. Now Jim "Hutch" Hutchison, mechanic, (See Chapter 20 for an outline of Hutchison's career) was to fly with Pat Reid with welding equipment to repair Gillam's crippled Stearman, still grounded at the *Nanuk*.

The two planes got off the next day. Matt Neiminen, an experienced Alaskan pilot, flew the other Fairchild 71, U.S. registry NC153H. He knew the route to Nome and led Pat Reid.

Snow was falling at Nome, and Seward Peninsula weather in general was bad. Neiminen lost sight of Reid's plane in a snowstorm, and returned to Nulato to wait the weather out. No one knew where Reid and the other Fairchild were.

Neiminen was stuck in Nulato, waiting for better weather. On January 6 he flew west from Nulato, searching for Reid's lost Fairchild without finding it. He returned to Nulato, and was again stuck there by terrible weather until January 10.

A dog team musher, searching for Reid, found the lost plane in a narrow bottom of the Ungalik River which drains into Norton Sound about 150 miles east of Nome. A wingtip was damaged, but Hutchison and a second mechanic aboard the plane were rebuilding it with wood from boxes aboard the plane, plus wood from local spruce trees. Fortunately, they had dope, fabric, and everything needed in the way of repair material. It was aboard the downed plane, intended for Gillam's Stearman.

Once repairs were completed, Reid took off and flew to Unalakleet, and then to Nome, arriving January 15. Permanent repairs to the new Fairchild's damaged wing were made there.

A letter from the Russian trader at Cape Serdtse-Kamen, carried by a Chukchi Native, arrived at the *Nanuk* on January 12, saying the missing Hamilton was heard flying west on November 9, and a few hours later it was heard flying east.

On January 21, 1930, Pat Reid and Ed Young, flying the two Fairchild 71 airplanes, arrived at Teller. Carrying gasoline and supplies for the *Nanuk*, they planned a short stop at Teller, and then the flight to North Cape.

Bob Gleason, on the *Nanuk*, received word that two planes from the ice breaker *Litke* were nearing North Cape, piloted by

Mavriki Slepnev and V. L. Galyshev, both of whom were experienced in the Arctic.

Powerful winter storms continued to sweep across Siberia, Bering Strait, and northwestern Alaska keeping all planes grounded for days on end.

Back at the *Nanuk,* Crosson managed a twenty-minute test flight of his Waco 10 on January 1, 1930. Storm after storm prevented him from flying for the next two weeks. It was bad enough that daylight hours were so limiting; the apparent impossibility of having enough light to fly and good weather simultaneously was frustrating. All Crosson could do was wait and guard his airplane from the deepening snow and high winds.

Tzaret Berdieff, searching with the *Nanuk's* dog team, had reports that the Hamilton was seen over one reindeer camp, inland twenty-five miles, and heard flying over another. Both were south of North Cape.

Crosson flew for more than two hours on January 17, checking out inland reindeer camps. He landed at three of them, without learning anything positive. On his return to the *Nanuk,* while searching nearby for the Hamilton, he took 17-year-old Marion Swenson up to allow her to see the sun, which was barely hidden below the earth's curvature. January 18 was when the first rays of the sun penetrated the winter darkness at North Cape.

Engineers from the Russian ship *Stavropol,* frozen in ice three miles from the *Nanuk*, repaired Gillam's Stearman; he didn't have to wait for Jim Hutchison to arrive with his welding equipment after all. Gillam flew his ship on a test flight January 20; skis and landing gear worked fine. Hutchison's trip to Siberia was cancelled.

For the next week falling snow and high winds kept Crosson and Gillam grounded.

Finally, January 26 dawned with a brilliant clear sky. The low-lying sun's rays lit up the land, throwing long shadows from planes, the *Nanuk*, and other objects. Crosson and Gillam were up early, warming their oil and engines.

They were off the ice at 9:30, headed about ninety miles for Cape Serdtse-Kamen. Word had come that the Hamilton had been heard there on November 9, flying east. They wore furs, as usual, in the wind-whipped open cockpits. Gillam flew a zigzag course two or three miles inland; Crosson flew parallel, but four or five miles farther inland.

The crashed Hamilton Metalplane NC10002 with Joe Crosson in the foreground. The up-ended wing of the downed plane threw a long shadow from the low-lying January Siberian sun, which caught Joe's eye as he and Harold Gillam searched for the missing plane. Harold Gillam

After about an hour, amidst many square miles of unbroken snow Crosson sighted something reflecting the sun, with a deep shadow beneath it. He banked the Waco and flew to check it out.

It was the left wing of the Hamilton, pointing forlornly at the sky. The fuselage was mostly buried by snow. It was about ten miles inland, and ninety miles from the *Nanuk*.

Gillam saw Crosson descend, and flew near. Crosson landed about 500 yards from the wreck. It was a rough area for landing, but usable, and he waved Gillam down. They walked across the hard-packed snow to the downed plane, and immediately realized it had hit the ground so violently there could have been no survivors.

The Hamilton had been headed southeast, probably at full speed, and it had hit a knoll. The right wing and the right landing

Fairchild 71 NC 154H landing at Fairbanks. This popular freight and passenger-hauling model bush plane first arrived in Alaska to help in the search for the lost Eielson and Borland. Eventually thirty of these planes flew Alaskan skies, some continuing into the 1950s.

gear were destroyed; the left wing lay at a forty-five degree angle, pointing skyward. The engine had been torn from the fuselage and lay buried by snow about 100 feet away. The prop was found wrapped around it later, when it was dug out. A few cylinders were found broken off. The tail and rear end of the fuselage had broken off.

The bodies of Eielson and Borland were not in, or visible around the plane. They had to be buried somewhere in the deep snow.

The food and case gas cargo was basically intact; the case of eggs barely damaged. When Crosson opened the cabin door a slab of bacon fell out.

The Hamilton's throttle was in wide open position. The altimeter read 1,000 feet. Point of impact was about fifty feet above sea level. The instrument panel clock, its glass cover broken, had stopped at 3:40; the plane had departed Teller at 10:45, indicating the flight had been underway for five hours less five minutes. Was the altimeter faulty? Was it reading properly when the plane crashed? Was it damaged in the crash? Was it stuck on the 1,000

PATRONS ARE REQUESTED TO FAVOR THE COMPANY BY CRITICISM AND SUGGESTION CONCERNING ITS SERVICE 1204

CLASS OF SERVICE
This is a full-rate Telegram or Cablegram unless its deferred character is indicated by a suitable sign above or preceding the address.

WESTERN UNION

NEWCOMB CARLTON, PRESIDENT J. C. WILLEVER, FIRST VICE-PRESIDENT

SIGNS
DL = Day Letter
NM = Night Message
NL = Night Letter
LCO = Deferred Cable
CLT = Cable Letter
WLT = Week-End Letter

The filing time as shown in the date line on full-rate telegrams and day letters, and the time of receipt at destination as shown on all messages, is STANDARD TIME.

Received at

RADI MS NANUK POINT BARROW ALASKA 1 25 30

EIELSON
HATTON N DAK

SORRY TO INFORM YOU FOUND BENS PLANE BADLY WRECKED ABOUT NINETY MILES EAST HERE STOP PLANE BURIED IN DEEP SNOW DID NOT LOCATE BODIES STOP VERY EVIDENT KILLED INSTANTLY STOP PILOTS COCKPIT TORN AWAY FROM ENGINE WHICH LIES ABOUT HUNDRED FEET FROM CABIN STOP DISPATCHING DOG TEAMS TO EXCAVATE SNOW AROUND PLANE

JOE CROSSEN

The telegram Joe Crosson sent to Ben Eielson's father Ole from the Nanuk via Point Barrow on January 25, 1930.

feet mark, or was it continuing to respond to the changes in atmospheric pressure after the crash? Was it set when the plane left Teller? These questions, of course, are unanswerable now, three quarters of a century later.

Crosson theorized that Eielson's first warning that his altimeter was wrong must have been just before he struck the mound. The throttle was wide open, he speculated, because Ben was attempting to climb over it. Normally, while cruising, the throttle would not be wide open.

Later, in a written report requested by the Soviets, Crosson said, "A faulty altimeter upon which Eielson was forced to rely because of poor visibility was the probable cause for the accident."

Pilots flying close to the ground, or while maneuvering, commonly keep one hand on the throttle, and the other on the wheel (the Hamilton had a dep[1] wheel, not a control stick). If Eielson was flying in that mode (with hand on the throttle), he could have jammed it wide open when the plane hit and abruptly stopped.

Why was the Hamilton found ten miles inland, when it was

1. Named for Deperdussin, the French aircraft builder who first used a wheel for an airplane's controls.

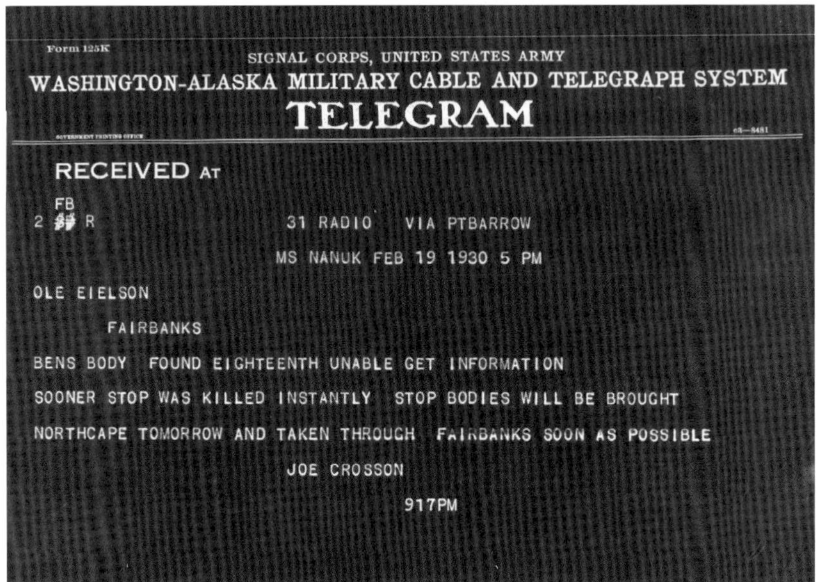
Form 125K

SIGNAL CORPS, UNITED STATES ARMY

WASHINGTON-ALASKA MILITARY CABLE AND TELEGRAPH SYSTEM

TELEGRAM

RECEIVED AT

FB

2 ## R 31 RADIO VIA PTBARROW

MS NANUK FEB 19 1930 5 PM

OLE EIELSON

FAIRBANKS

BENS BODY FOUND EIGHTEENTH UNABLE GET INFORMATION SOONER STOP WAS KILLED INSTANTLY STOP BODIES WILL BE BROUGHT NORTHCAPE TOMORROW AND TAKEN THROUGH FAIRBANKS SOON AS POSSIBLE

JOE CROSSON

917PM

The February 19, 1930, telegram from Joe Crosson to Ben Eielson's father.

bound for the *Nanuk,* frozen in ice offshore in the Chukchi Sea? Snow had been falling for a week at the *Nanuk* as well as on the Alaska side of the Bering Strait. Sea ice and the land on the Siberian side may have presented Eielson nothing but deep snow unsuitable for a wheel landing. It can only be theorized, but Eielson, caught in a snowstorm with limited visibility, may have flown inland looking for a safe place to land his wheel-equipped airplane.

On November 9th, date of the crash, light from a very low-lying sun would have been dim at 3:40 p.m. The sun disappeared at North Cape on November 15 and was not seen again for two months. Add heavy falling snow or fog to dim light, and the Hamilton could have been flying in a whiteout. Whiteouts are not uncommon in the Far North.

Crosson and Gillam dug dispiritedly around the plane, looking for the bodies. They had no digging tools, and their hearts weren't in it. The bodies had to be somewhere beneath the deep snow, which was hard-packed as the result of many weeks of high wind. Shovels, and perhaps axes, were needed.

Crosson loaded into the front cockpit of his Waco the case of eggs, some bacon, and a piece of aluminum from the crashed plane. He and Gillam made the hour-long flight back to the *Nanuk.*

The *Nanuk's* crew gathered around as they taxied to a stop. Crosson wearily climbed out of his Waco and quietly announced, "The search is over. We've found them." He then showed them the piece of corrugated aluminum from the crashed plane.

13 Recovering the Bodies

THE ARCTIC DID NOT give up its secrets easily. Seventy-seven days passed from the time Eielson and Borland left Teller until they were found. Pilots who wanted to search for the lost Hamilton spent most of that time frustrated, waiting for breaks in the weather so they could get on with the search. They endured constant zero-visibility fog, deep cold, snowstorms, high winds, short daylight hours, and broken airplanes.

Now the bodies had to be recovered from an extremely remote region in a foreign land, and from almost cement-hard, packed snow. It was not to be a simple task. The bodies had to be returned to Alaska and taken on to the states for burial. Airplanes needed to be returned to Alaska. Gasoline had to be flown to North Cape to refuel planes, and to replace borrowed supplies. There was still the fur to be flown out, as well as people.

AN INTERNATIONAL GATHERING

On January 28, with good weather at Teller as well as North Cape, two Fairchild 71 airplanes arrived at the *Nanuk.* Ed Young flew U.S. registered NC153H, and Pat Reid, flew Canadian registry CF-AJK. Now there were four planes based at the *Nanuk*, the Waco 10 of Joe Crosson, the Stearman flown by Harold Gillam, plus the two Fairchilds.

On the next day, two German-designed low-wing monoplane Junkers from the ice breaker *Litke* at Providence Bay arrived. On

Pilots who participated in the challenge of Siberia (l. to r.) were Pat Reid, Canadian volunteer, Ed Young, who flew the funeral Fairchild NC153H from North Cape to Fairbanks, Harold Gillam and Joe Crosson, who found the downed Hamilton. All wear the furs they needed to function in frigid Siberia during the search and body recovery.

skis, they circled high over the *Stavropol*, and descended to land on the strip near the *Nanuk*, where the other planes were parked.

One landed downwind, fast. It touched down on a smooth part of the runway, but skidded at high speed into hard-packed snowdrifts at the runway's end. At the first of these it bounced high, to come down with a loud thump, only to be re-launched into the air by another hard drift. After several bounces that carried it ten or more feet high, it landed hard, breaking one of the landing struts. The right wing ended in the snow, supporting that side of the plane.

The second Junkers pilot, perhaps learning from his companion's experience, landed upwind. He might have been playing, most likely not; at any rate he plunged out of the sky and it appeared he was going to land on the Fairchilds and the planes of Crosson and Gillam. As he neared the bow of the *Nanuk*, the gathered ship's crew and the four pilots, fearing the worst, frantically scattered.

The Junker skimmed a few feet over the four parked planes and for a moment it appeared one wing was going to hit the other planes. But the pilot used full throttle and climbed out, circled, and tried again. This time he touched down about three quarters of the way down the marked runway. He too skidded off the runway into rough snow, but nothing broke.

Fortunately, that plane had spare parts to repair the first plane's landing strut.

The F-13 Junkers low-winged monoplanes were probably built in Russia from German plans. They had closed cabins for passengers, but the pilot and mechanic rode in an open cockpit behind a windshield near the front of the plane. The liquid-cooled L-5 inline German-designed engines were rated at 310 horsepower.

Pilot Mavriki Slepnyov and mechanic Fabio Fahrig manned the first Junkers to land; Victor L. Galishev piloted the second, with mechanic Brednevya.

Now, six airplanes and pilots from three nations were gathered at the *Nanuk*. The original assignment for the two Junkers was to fly passengers to Providence Bay from the nearby ice-locked *Stavropol*. However, they soon received orders to first help recover the bodies of Eielson and Borland.

The Digging Starts

Joe Crosson was in charge of American digging efforts. Weather permitting, dog teams and airplanes started hauling volunteers, food, camping gear, and digging tools the ninety miles from the *Nanuk* to the wreck site. Crosson, Gillam, Reid, and Young, flew almost constantly between the *Nanuk* and the wreck. Their planes gulped gasoline until the supply was nearly gone. At first there were only a few workers with shovels at the wreck, and it was assumed the job to find the bodies would take but a day or two. This proved wrong. It took three weeks to find both bodies.

Pat Reid was assigned to fly Fairchild 71 CF-AJK to Teller for a load of gasoline, but he had to wait until February 7 for suitable flying weather. Olaf and Marion Swenson accompanied him on the flight to Alaska, as did the captain of the *Stavropol*, who needed medical attention. The Swensons continued on to Seattle.

In his fine book, *Icebound in the Siberian Arctic*, Bob Gleason commented that he and Marion Swenson had been the two youngest aboard the *Nanuk.* They had been drawn together, and had become quite fond of each other. Under the crowded living conditions, and with the ever watchful father Olaf, the situation wasn't appropriate for romance, he said. He wrote of Marion as "… a brave, beautiful, young woman."

The Swensons arrived in Seattle March 1. Olaf immediately bought the Danish motorship *Karise* to replace the lost Elisif. By June, 1930, he was en route to North Cape with it, loaded with trade goods and supplies, as he had been with the *Nanuk* a year earlier.

Joe Crosson, with the Waco biplane, flew Commander Slepnyov to the wreck site on January 30. As usual Joe made his landing the American and Alaskan way—coming in just above stalling speed for a short landing run. The Russian pilot was amazed. Pilots in their service were taught to land with plenty of speed to keep good control. This, of course, resulted in a much longer run after touching down, and the need for a longer air strip. This explained their slap-dash landing upon their arrival at the *Nanuk.*

Slepnyov received radioed orders at the *Stavropol* to fly a Junkers to the wreck and to take charge of the Russians (crew of *Stavropol*) who were to search for the bodies. He responded that he didn't think he could land a Junkers there without wrecking it.

"Land anyway," he was ordered.

CAUSE OF THE CRASH[1]

While waiting for flying weather, Slepnyof questioned Crosson, Gillam, Young, and Reid for their views on the cause of the accident. Olaf Swenson translated. There was general agreement that Eielson had found himself in near-darkness, without adequate flight instruments for whiteout conditions, and had unknowingly flown into rising ground perhaps while looking for a landing place.

The Canadian pilot Pat Reid also theorized that Eielson may have

1. A 2004 study found that the average life expectancy of a non-instrument-rated pilot who flies into clouds or instrument conditions is 178 seconds. None of the references I found on Eielson indicated he had any special training for instrument flying, although he was reported several times as "going on instruments" while flying with Wilkins. In 1929, instrument flying, as we know it today, did not exist. Jimmy Doolittle made the first take off, flight, and landing solely on instruments on September 24, 1929. It was years before true instrument flying was common. JR

Pilots and others involved in the Siberian challenge gather at the University of Alaska in March, 1930. FROM LEFT, FRONT ROW: Peter Nickoloff, Russian translator; Joe Crosson, pilot; Matt Neimenan, pilot; Bill Hughes, Canadian mechanic; Sam Macauley, Canadian pilot. SECOND ROW: Fahrig, Russian mechanic; unidentified; Commander Slepnyov, Russian pilot. BACK ROW: Harold Gillam, pilot, Fairbanks Mayor De la Vargne; Dr. Charles Bunnell, President of the University of Alaska; and Otto Geist, German translator and archaeologist/palaeontologist at the University.

been caught in a whiteout, when snow-covered ground and heavy falling snow blend; everything appears white, there is no horizon, no sky, no ground visible. It has been compared to flying inside milk.

Weather was stormy February 2, 3 and 4. All planes were grounded.

On February 5, after the site was selected by Joe Crosson and Harold Gillam, Slepnyov successfully landed one of the Junkers about 350 feet from the wrecked Hamilton. With blocks of snow, workers built a shelter under one of the Junkers' wings.

The Junkers' cabin and its under-wing snow shelter, plus two tents, heated with wood stoves, were used as a refuge for diggers, and in more-or-less moderate temperatures, as rather harsh overnight shelters. They had first spent nights at the six-mile-distant cabin of Brokhanov, a Russian trapper. Wood for the tent stoves was hauled by dog teams from Brokhanov's cabin. When the temperature plunged, all the men, with dog teams made a nightly trek to Brokhanov's cabin, which, though crowded, was much warmer.

Soon, fifteen (eventually nineteen) men from the *Stavropol* started digging trenches through the deep, hard-packed snow. Parts of the Hamilton were scattered over several hundred square feet, and in time, trenches radiated many directions from the main wreck.

The first finds were part of the plane's cabin, a landing wheel, and Borland's helmet, broken goggles, and mittens, which he probably had not been wearing in the heated cabin before the crash. Next, scattered widely, were found miscellaneous tools and seat cushions.

Trapper Brokhanov helped. At one point, he urged the shovelers to dig beneath wolf tracks he had seen, thinking the wolf might have been attracted to the bodies. The slow toil went on for days, occasionally interrupted by violent snowstorms and, once by deep cold and heavy snow.

Borland Found

The diggers were up at 7:30 on February 16. They had tea and a little food as usual and returned to their work. Several hours later Russian worker T. Jakobson found the body of Earl Borland. It was under five feet of hard-packed snow, and about four feet from the Hamilton's engine. The body, face down, was wearing a parka, the head was toward the wreck. So hard was the snow that it took until after dark before the workers could chip enough of it away to remove the body. When dark arrived, the body was covered and left near where it was found.

A blizzard arrived the next day, February 17, with visibility in mere feet. Digging was impossible. The temperature plummeted to -20 F. Snow drifted to the top of the parked Junkers' fuselage, and filled all the newly dug trenches. The two tents, one 8 x 10, the other 10 x 12, were pitched near each other. Four workers remained in one of the tents overnight; all others returned to Brokhanov's cabin. It was at least warm.

On the morning of February 18 the four workers in the tent found the flaps frozen shut, and the tent itself nearly buried in snow. They had to dig their way out. Though the tent was heated by a wood stove, it was so cold that when they drank hot tea, they needed gloves to keep their fingers warm.

That afternoon when several men tried to travel through the blizzard 350 feet from their tent to retrieve firewood stored by the Junker airplane, they lost their way despite a row of stakes driven into the snow as guides. They had to return to their tent. They managed to get the firewood on a second try.

The storm continued. Huge drifts of snow built around the tents. A strong wind drifted snow, and it wasn't possible to dig for Eielson's body. Borland's body had been left out, though covered by canvas.The wind dropped off that afternoon, and the crew again started digging. One of the first goals was to dig the body of Earl Borland from newly drifted snow.

Harold Gillam with Brednevya, one of the Russian mechanics, as his passenger, landed nearby with supplies from the *Nanuk.* As Harold climbed out of the Stearman, he saw the workers lift Borland's body from the snow. He stared for a moment, and returned to lean against his plane, clearly overcome. The body was contorted, and frozen solid.

The Russian workers carried the body to the parked Junkers, carefully wrapped it in a tent, and placed it in the airplane's cabin. Slepnyov locked the door of the Junkers and walked to Gillam, who was still badly shaken. He could hardly speak. "We appreciate all the work you and these men are doing," is all he could get out, his voice breaking.

By a quirk of fate, Gillam was flying the Wien Stearman biplane that Earl Borland had spent two weeks repairing at Walker Lake the previous April.

Snow was now twice as deep as when they had first found Borland's body, increasing the difficulty in finding Eielson's body.

Eielson Found

Finally, on February 18, the body of Ben Eielson was found by volunteer digger T. Jacobsen, who had also found the body of Earl Borland. It was face down and lightly covered with snow, about 200 feet from the wrecked plane, and 150 feet from where Borland's body had been.

The Russian Junkers F-13, flown by pilot Mavriki Slepnyov, was at North Cape to assist in the Eielson Search. In it Slepnyov flew the bodies from the wrecked Hamilton to the Nanuk. It was one of the three "funeral planes" that accompanied Ed Young's Fairchild 71 NC153H from North Cape to Fairbanks.

This photo was taken while the plane was at Fairbanks in March, 1930. CCCP-177 was dismantled at Fairbanks and sent back to Russia by rail and ship. George Bishop

Workers wrapped it in a tent and stored it next to Borland's body in the cabin of the parked Junkers.

Clearly, both airmen had been killed instantly when the Hamilton crashed.

The temperature dropped to -30 F. next day as the crew prepared to leave the wreck site. The Russian mechanic Fahrig started work to start the Junkers engine. Another blizzard arrived, which ended work on the airplane for that day.

On February 20 Fahrig was again trying to start the Junkers when Harold Gillam landed with the Stearman. He learned that Eielson's body had been found and was in the cabin of the Junkers with that of Borland. He asked if he could look at Ben's remains, and was led to the Junkers.

Gillam climbed into the airplane cabin and stood quietly looking at the bodies of his two friends. He had worked with

Borland when both worked for the Alaska Road Commission, and more recently when both had worked as aviation mechanics; Eielson had encouraged him to learn how to fly. The Fairbanks circle of pilots and airplane mechanics was small and close. Most Fairbanksans had known friendly Ben Eielson, as a teacher and as a pilot. His international fame hadn't affected him; he was still good old Ben, the former high school teacher, and a local pilot who had made good.

Gillam left the Junkers and watched Fahrig and one other mechanic trying to start the Junkers' engine. The fuel line was partially plugged, and the engine refused to start.

Gillam loaned his blowtorch to Fahrig and showed him how to use it. After waiting some time for the Junkers engine to start, without results, pilot Slepnyov decided to fly back to the *Nanuk* with Gillam so he could officially report from the *Stavropol* on the finding of both bodies.

February 21 was stormy. On the 22nd, temperatures plunged to -50 F.

Pat Reid Breaks a Fairchild

On February 22, with the bodies aboard and the Junkers CCCP-177 engine running, Slepnyov was preparing to take off when Pat Reid arrived overhead from Teller, flying Fairchild 71 CF-AJK with a load of cased gasoline. Reid saw the Junkers and Gillam's Stearman. Smoke rose nearby either from one of the tents, or a warming fire. He assumed it was a signal and decided to land.

It was a mistake that ended in disaster.

On the ground, the Fairchild hit a solid ridge of wind-packed snow. This broke a ski, which flipped into one of the cabin's windows. The landing gear was torn from the plane. The prop was curled when it struck the hard-packed snow. It was his second accident with the new Fairchild, and he was deeply chagrinned. Fortunately, he was uninjured.

The Fairchild would remain next to the remains of the wrecked Hamilton until Jim Hutchison and his welder could be flown to it from Fairbanks for final repairs.

Slepnyov flew the Junkers CCCP 177 with the bodies to the *Nanuk*. As the plane touched down, flags on the *Nanuk* and the *Stavropol* were at half-mast. Local villagers, people from the Stavropol, and the crew of the *Nanuk* were gathered to meet the Junkers.

Ceremonies of Respect

The bodies were placed in the only frame building at North Cape to thaw so they could be placed in a reclining position. The building was draped with American and Russian flags. A twenty-four hour watch by two guards – one American, one Russian - was kept over the bodies

When the bodies were thawed and straightened, they were taken to the *Stavropol* where Doctor M.V. Kreszanev, of the ship's crew, could examine them for his official reports.

After his examination, the bodies were sewn into white drill cloth. Chukchi women from the village of North Cape had created American flags from strips of red, white, and blue muslin sewed to white ship's canvas. These were draped over the drill-cloth-covered bodies.

On February 26 the bodies were placed on two Siberian dog sleds. With Ben's and Earl's faces exposed, the sleds traveled between two lines; one of Americans and Canadians, the other of Russians and local Natives.

They arrived at Ed Young's Fairchild 71 (NC153H) that was to fly them to Alaska. At the Fairchild, Commander Mavriki Slepnyov and the acting captain of the *Stavropol* presented the bodies to pilots Ed Young, Joe Crosson, and Harold Gillam, who accepted them on behalf of the Americans.

The leader of the Cuckolk district, in which the Hamilton had crashed, had traveled by dog team 400 miles to be included in the ceremony. He spoke briefly, saying how sad he was that the tragic accident had happened on Soviet land.

All stood at attention as the bodies were loaded into the Fairchild. Except for the cockpit, all its windows were shrouded in black.

The funeral plane, accompanied by Commander Slepnyov and his mechanic in Junkers CCCP 177, landed at Teller and refueled. Gillam and Pat Reid had earlier flown to Teller with the Stearman, and were there to meet the funeral plane.

The three planes, in formation, overflew the Nome airport

because of windrows of hard-packed snow there from a recent storm. Instead, they landed at Ruby.

Delayed by weather, the three planes finally arrived at Fairbanks March 7, flying in formation, the funeral plane in the center, the Junkers on one side, the Stearman on the other. Aboard the funeral plane, flown by Ed Young, were Joe Crosson and Canadian mechanic Sam McCauley; Gillam's Stearman carried Pat Reid and Canadian mechanic Bill Hughes. Near Fairbanks, the funeral plane pulled ahead, and the other two planes followed single file. They circled the field once and landed. A large crowd had gathered.

Young taxied the funeral plane to a hanger, where it was pushed inside and the hangar doors closed before the bodies were removed.

Ben's father, Ole, and Earl Borland's father William, had arrived together at Fairbanks to await the funeral plane. When Ole arrived at his hotel he was handed a sympathetic telegram from President Herbert Hoover.

In Fairbanks, services for the two lost aviators were conducted jointly at the Moose Auditorium by the American Legion and the Pioneers of Alaska. Most residents were there. Memorial services were also held for the two at Teller, Nome, Ruby, and Barrow.

While in Fairbanks, the Russians presented medals to Joe Crosson, Harold Gillam, and Ed Young. They were mounted on a silver plaque, with all three of their names included, and the words, "To brothers and friends, from Slepnyov."

The Bodies Leave Alaska

The bodies traveled to Seward on the Alaska Railroad, and from there to Seattle aboard the steamer *Alaska*, which arrived after dark. Seattle residents were alerted to the arrival by the sound of the *Seattle Times* whistle.

The Seattle funeral cortege for Ben Eielson attracted throngs. His flag-draped coffin was placed on a military caisson that took it to the railway station. In the procession were the Russian pilot Slepnyov and mechanic Fahrig, Canadians Reid and mechanic Hughes, North Dakota National Guard units, members of the Seattle Police Department, United States military representatives, as well as American Legion members, trader Olaf Swenson and

Sunday, March 23, 1930, Seattle, Washington. The caisson bearing Ben Eielson's body was led by police and highway patrolmen, and escorted by North Dakota National Guardsmen, as well as American, Canadian, and Russian pilots and mechanics who participated in the search for the lost Hamilton. Hundreds crowded the walks to pay tribute as the body was taken to the railway station for transfer to a private Pullman car which carried Eielson to his home town of Hatton, North Dakota, for burial.

daughter Marion, Alaska-Yukon Pioneers, and representatives of the Sons of Norway. As it wended its way through Seattle huge, silent, and solemn crowds crowded the sidewalks and streets.

Alaska-Yukon Pioneers conducted services for Earl Borland, who was buried at Seattle. He was nearly 30, and left his wife Irene and young sons William "Bill" and Earl Jr. (Bud). Bill was a U.S. Marine pilot during World War II, and later became Chief Pilot for Alaska's Reeve Aleutian Airlines.

The year 1929 was a time of adulation of heroes, and Ben Eielson was the perfect hero. He was modest, from ordinary folk, and, as the world had watched, he had searched for new lands from the air—a new, exciting, and hazardous undertaking. Today, it sounds a bit odd to talk of searching for new lands, but that romantic goal provoked much excitement at the time.

Eielson was 32 at the time of his death. His coffin traveled in state from Seattle aboard a private Pullman car (the same car used by Amundsen to travel east after his polar flight). On its way to his home town of Hatton North Dakota it was covered with flowers, and more arrived at every stop. In Montana and North Dakota it stopped at small towns to allow residents a moment with their hero. Airplanes, including a Lockheed Vega, flew over and escorted the train between stops. A huge crowd met the train at Minot, North Dakota.

There was no television of course. Commercial radio stations were just becoming established. News came mostly from newspapers. Today, with up-to-the-minute news arriving by television, it is unlikely that such crowds would gather to watch a funeral train pass, or so many would congregate in a small town for a funeral; instead, most viewers would participate by watching the tube.

A crowd estimated at 5,000 gathered at Hatton, a small Midwest town in northeastern North Dakota. Sir Hubert Wilkins and aircraft mechanic Orville Porter flew there to attend the ceremonies. An overflow crowd of more than 4,000 stood outside the little St. John's church during the funeral service. North Dakota's governor George Shafer, among others, spoke. Prominent citizens praised Ben's achievements. Ben's family received many sympathetic telegrams from the U.S. and around the world.

A thousand mourners walked the two miles from the church to the local cemetery where Ben was laid to rest beside his mother and a brother.

Finishing the Job

Fur remained on the *Nanuk*, waiting to be flown out. The Waco 10 (NC780C) that Crosson had flown was still parked next to the *Nanuk*. And the Fairchild 71 (CF-AJK) that Pat Reid had broken while loaded with gas was still lying on its belly at the Hamilton crash site.

Mechanic Herb Larison had been left on the Siberian side to repair the Fairchild 71. In ten days he dug a hole in the snow under the plane and did what he could to repair the landing gear. He then returned ninety miles by dog team to the *Nanuk* to have

Bob Gleason radio Fairbanks the need for Jim Hutchison and his welding equipment to complete repairs.

On April 26, 1930, Joe Crosson flew Fairchild 71 NC9153 from Teller to the Hamilton wreck site. With him were Jim Hutchison and pilot S.E. Robbins. Hutchison did the required welding, and on the next day Crosson flew the repaired Fairchild CF-AJK to the *Nanuk.* S.E. Robbins followed with the Fairchild 71 NC9153.

With the bodies found and flown back to Alaska, there was still the baled fur to be flown to Fairbanks. There was some urgency for this, for the fur market was falling; the great depression of the 1930s had begun to show its teeth.

There was still about 10,000 pounds of valuable fur on the Nanuk. Fog held the aviators at the *Nanuk* until May 6, when both Fairchilds were flown to Teller with forty sacks of fur. Next day they flew to Nome, where they unloaded the furs and turned them over to the United States Post office, which assumed responsibility for flying it on to Fairbanks.

Joe Crosson, with S.E. Robbins, who was to fly the Waco 10 back to Alaska from the *Nanuk*, and mechanic Herb Larison, flew from Teller to the *Nanuk* on May 8. Both planes returned to Teller on May 10, with Larison riding with Robbins in the Waco, and Joe flying Fairchild NC9153 with twenty-six bags of fur.

All Alaskan Airways planes were now out of Siberia except for the crashed Hamilton[2]. The valuable fur—ermine, fox, and marten - had been removed from the *Nanuk*, leaving less valuable squirrel, wolf, seal, and other miscellaneous furs behind. The fur-hauling contract was completed.

The cost? The loss of Eielson and Borland. No figure could be set.

As for the rest, there was the loss of the $26,000 Hamilton airplane, the dollars spent for the extra flying, repair of airplanes, and the additional hundreds of gallons of fuel used in the search and its aftermath.

Gained: Invaluable experience in deep cold flying for a fledgling Alaska aviation industry.

The search for Ben Eielson and Earl Borland was the greatest challenge of the 1920s to Alaska's pilots, aviation mechanics, and

2. Soviet historians salvaged the remnants of Eielson's lost Hamilton, and a Soviet airplane flew them to Fairbanks on March 5, 1991. They are on display at the Pioneer Air Museum in Fairbanks.

Eielson Air Force Base is but one of many establishments or places that honor the name Eielson in Alaska.

airplanes. Through determination and skill, a handful of dedicated professional airmen found their lost brothers in as near an impossible situation as could be imagined. They also completed a contract—flying the *Nanuk's* fur out—a superb aerial achievement for the time.

The dangers of the Siberian challenge to Alaska's airmen cannot be overstated. Pilots and mechanics risked their lives with every flight; there was always a substantial chance of being forced down on the ice of the Bering Sea while making the five-or-six-hour-or-so jump from Alaska to the *Nanuk.*

It was truly an heroic effort.

Eielson Memorials

Alaskans live with the name Eielson; there is Eielson Air Force Base, a home to modern jet fighters and other military planes, twenty-six miles southeast of Fairbanks adjacent to the Richardson Highway. A monument there was dedicated to Ben's memory on July 20, 1956. On it are scenes of Alaska, an outline of his life, and a list of his contributions to aviation.

There is the Eielson Memorial Building on the campus of the University of Alaska Fairbanks. There is even a block-and-a-half-long Eielson Street in Anchorage, near the junction of International Airport Road and Minnesota Drive.

Mount Eielson, in the Alaska Range, is 5,802 feet high, in Denali National Park. It lies three and a half miles south of the park's Eielson Visitor Center. Before it became Mount Eielson it was known as Copper Mountain. It was re-named Mount Eielson on June 14, 1930 by the U.S. Congress for Carl Ben Eielson ..."in honor of his pioneer work in aviation performed in Alaska and the North."

This action was taken a scant four and a half months after his body was found. Has any other action by the Congress in recent years been as prompt? Like Lindbergh, Eielson was an international hero; honoring him by naming a small mountain for him was a slam-dunk for Congress.

There is also Eielson Peninsula, in Antarctica, named by Sir Hubert Wilkins.

In 2003, Ben Eielson was inducted into the Alaska Aviation Heritage Museum Hall of Fame.

In 2004, Fairbanks Chapter 1129 of the Experimental Aircraft Association restored to airworthiness the Curtiss JN-4D airplane

The display at the Fairbanks' Pioneer Air Museum of the Hamilton Metalplane's remnants returned to Alaska by Russian historians. The wing backing the display was the left wing of the wrecked plane pointing toward the sky that attracted Joe Crosson's attention as he flew a search for the plane. Author that in 1923 Eielson flew from Fairbanks, and that others, including Joe Crosson and Noel Wien, flew there in following years. That Jenny once hung from the ceiling at the Fairbanks' International airport terminal. However, its wings were substitutes from another airplane. The originals, hung from a hangar wall at Weeks Field, were lost when the hangar burned.

The restoration work, including the wings, which are now built to Jenny specifications, took place in the Hutchison Career Center's aviation shop. The airplane (it's photo is on the front cover of this volume) now hangs above the baggage carousel in the new Fairbanks air terminal. Local businesses donated more than $20,000 for the restoration, which was accomplished by more than thirty-three volunteers.

The frame house in Hatton where Ben and his eight siblings were raised is now the Hatton-Eielson museum dedicated to Ben's memory. Within it is a library, trophies, photos, and souvenirs, and a wealth of memorabilia on Eielson and his historic flying career.

Several North Dakota Schools have been named for Ben Eielson.

In 1956, an Eielson Memorial Committee, formed after Ben's death, dedicated the Ben Eielson Memorial Arch which is now at the main entrance of St. John's cemetery a mile or two north of Hatton. His name is carved on the upper part of the arch; a bronze plaque with Ben's likeness is mounted on one base; an inscription recounting his life is mounted on the other.

14
Payback Time

THE SINKING OF THE *CHELYUSKIN*

In March, 1934, the Russian ship *Chelyuskin*, 150 miles north of Teller, Alaska, and seventy miles from Siberia, was crushed by ice, and sank within two hours, stranding 104 Russians on a big ice floe. They needed immediate rescue. Led by famed Soviet Arctic explorer, Dr. Otto Schmidt, the vessel was on a scientific expedition seeking a northern sea route, plus it had been transporting young families to a settlement on Wrangel Island.

Less than a week after the sinking, Russian pilots Sigismund Levanevsky and Mavriki Slepnyov, accompanied by Dr. George Ushakoff, Russia's rescue team leader, arrived at Fairbanks to purchase two airplanes to use for the rescue, and to hire two American mechanics to care for them. This was likely a result of Slepnyov's experience with Americans, and their airplanes' performance when he participated in the retrieval of bodies in the aftermath of the Eielson/Borland search.

The airplanes, purchased from Pacific Alaska Airways, were nine-place Consolidated Fleetsters, with 500-horsepower engines.

The *Chelyuskin* survivors had set up camp with what they had managed to toss from the sinking ship. Ice conditions prevented use of rescue ships, leaving aircraft as the only possible way to get the stranded people to safety.

Clyde Armitstead, then 34, a licensed aircraft mechanic in Fairbanks, was a partner in the most complete, licensed airmotive shop in the Territory. He knew and worked on the planes of virtually every pilot who flew out of Fairbanks. One day in early

March, 1934, he received a call from the shop foreman of Pacific Alaska Airways. "Some Russians at the Nordale Hotel want to talk to you about a job," He said.

"I don't want to go anywhere," Armitstead told him.

"Go talk to them anyway," he was urged.

By the time he had finished talking with the Russians, Armitstead had signed a contract to go to the aid of the *Chelyuskin* survivors as an aircraft mechanic. He had also agreed to find one other mechanic who would go. After a search, the only one he could find who was willing to go was 20-year-old Bill Lavery, born and raised in Fairbanks, a licensed pilot with an air transport rating who knew how to care for airplanes in deep cold. Lavery had soloed at the age of 16 at the Curtiss-Wright Flying School at San Mateo, California. The Russians accepted him as a mechanic with Armitstead's strong recommendation.

Mechanic Jim Hutchison commented later that he thought the Russians wanted American mechanics because they didn't know how to read the flight instruments on the Fleetsters. He thought they had no idea what RPM meant, or, for that matter what the readings on other flight gauges signified.

The four men flew to Nome in the Fleetsters, Armitstead with Levanevsky, and Lavery with Slepnyov. From Nome they flew across Bering Strait. Levanevsky's ski-equipped Fleetster iced-up, and he crash-landed on the ice of the Chukchi Sea about a mile offshore of Siberia. A local Eskimo rescued the pilot and Armitstead, and took them ashore with his dog team, to spend the night in his driftwood hut.

Slepnyev rescued Armitstead and Levanevsky the next day with the other Fleetster. The wrecked Fleetster was abandoned, and the surviving Fleetster flew the four men to the drifting ice floe and the stranded Russians.

Over the next five weeks the surviving Fleetster and five German-made biplanes ferried all of the people from the ice floe to the mainland. They had been stranded on the ice for two months; three babies were born while the party was on the ice floe.

Upon completion of the rescue, the Soviet government paid Armitstead and Lavery each $2,500 (a year's wages for many Alaskans at the time), and gave them an expense-paid trip to Moscow where they were feted as heroes.

While they were in Moscow in April, 1934, at a banquet room

in the Kremlin, A.S. Yenukidze, Secretary of the Central Executive Committee of the U.S.S.R., pinned the Order of Lenin on the mechanics' chests, and declared, "It is with supreme joy we declare that the heroic expedition terminated in victory."

The medals, ordered by Stalin, came with thirty rubles ($25 at the time) a month for life, plus free transportation for life by rail or boat in the U.S.S.R. Only pregnant women and recipients of the medal were afforded these travel privileges.

The thirty rubles stopped arriving in December, 1947, when the Western allies organized the Berlin airlift. For life, the rubles weren't.

Three years after the rescue, Armistead flew as an observer on another search, when Sigismund Levanevsky's four-engine airplane disappeared on a flight from Moscow to Fairbanks (see following).

At the age of 7, Clyde Armitstead crossed Oklahoma with his family in a covered wagon bound for New Mexico to homestead. He started his aviation career in 1923 with the Marine Corps, and first worked on Jennys. He worked at the Wright Engine Factory during World War II, and taught aviation to civilian women; he was too young to serve in WWI, and too valuable as an instructor to serve in WWII.

He left Alaska when World War II started, returned in 1969 after retirement. He died at age 94 on February 11, 1997, at Eagle River, Alaska.

For seven years, William (Bill) Lavery flew for his own Lavery Airways, at Fairbanks, which he founded at age 20. He later flew for Pan American Airways, Northwest Airlines, Wien Airways, and Alaska Airlines. He was the manager of Fairbanks International Airport 1973-75, then retired to his cabin on the Goodpastor River. He died in Fairbanks on December 10, 1983, at age 69.

The Loss of Sigismund Levanevsky

Flight Across the Pole

On August 17, 1937, a big blue and red, four-engine ANT-6 airplane, Soviet registration N-209, flown by famed Russian Arctic pilot Sigismund Levanevsky, (who, with Mavriki Slepnyov, had convinced Armitstead and Lavery to help in the rescue of the marooned *Chelyuskin* passengers), climbed into the sky from Moscow and headed north. His goal was to make a non-stop flight across the North Pole to Fairbanks, Alaska. Levanevsky, sometimes called "The Lindbergh of the North," was accompanied by a crew of five. Aboard the plane was a complementary cargo of mail, furs, and caviar. The flight was one of three the Soviets made that year in an attempt to establish a commercial polar transport route between the Soviet Union and North America.

Levanevsky didn't make it. Residents of Fairbanks, and Soviet officials waited for the plane at Weeks Field. The disappointed crowd dispersed long hours after the plane should have arrived.

At the time, Levanevsky was famed for round-the-world flights, as well as multiple flights in the Arctic. In 1933 he piloted the Soviet airplane that ferried failed American round-the-world pilot Jimmy Mattern from Anadyr, Siberia, to Nome. Mattern's Lockheed Vega had crashed on the tundra when he attempted to fly from Khabarovsk to Nome.

During their over-the-pole flight, the position of N-209 was reported by the plane's radioman, Nicoli Galkovsky, as being on the 148th meridian and half way (roughly 300 miles) between the North Pole and Alaska's arctic coast. He also reported, "Continuous clouds and a sixty mph headwind."

His last report, "Outside right engine has failed on account of damaged oil pipe. Altitude 15,000 feet. Descending to 13,000 feet. Flying through solid cloud. Stand by," left the world speculating.

Was something beside a 'damaged oil pipe' ongoing with the airplane that prompted the "Stand by?" Also, the report of sixty-mile-per-hour headwind caused speculation that the plane may have run out of fuel.

The Search for the N-209

To search for the lost N-209, the Soviet ice-breaker *Krasin* soon arrived at Barrow, with fliers and a Fokker airplane. The Soviet government asked for American aid.

Sir Hubert Wilkins became the Commander of the Alaska-Canada section of the contingent searching for Levanevsky. The search lasted for nine months, from August 1937 into the spring of 1938.

Wilkins flew a Consolidated PBY on the search (a twin engine flying boat with a range of 2,000 miles, a type that became famous in World War II). He even flew on clear moonlit nights, when visibility was nearly as good as daytime. He flew searches for up to 1,000 miles a day for the first month, and continued into winter as long as there was hope the lost Russians could still be alive.

Joe Crosson and Harold Gillam, recalling the all-out help and cooperation of the Russians with the Eielson/Borland search, realized it was payback time. Both pitched in to help. Crosson spent weeks that August and September flying great distances over the ice north of Barrow, looking for the N-209, as did the Russian Fokker and Canadian pilots.

No sign of the airplane or its crew has ever been found.

By 1931, Harold Gillam had established his own flying service, although he had a rocky start by breaking six airplanes in the first six months of his new business. Now he used his workhorse American Pilgrim on many support flights to Barrow for the Soviets. He was building a reputation as a pilot who commonly flew in weather that kept other pilots on the ground. "He flew in weather when even the ravens were walking," other pilots commented.

His American Pilgrim was capable of flying for seven hours on the fuel it carried, and under good conditions the flight from Fairbanks to Barrow required six and a half hours. That's cutting it close.

Dan Cathcart, a United Airlines pilot, who later worked for a short time as a pilot for Gillam, accompanied Gillam on a flight from Fairbanks to Barrow with supplies for the Soviet searchers. He reported that Gillam climbed high to clear the Brooks Range, and after that he flew above clouds toward Barrow for hours with no view of the ground.

Gillam nosed the ship down through the blinding mist. The first thing Cathcart saw as they landed on the Barrow lagoon were the houses of Barrow and antenna poles about 500 feet from the lagoon.

"I don't know how he did it," Cathcart said.

"Harold Gillam wore the Fairbanks-Barrow air route smooth, freighting aviation gasoline and supplies to the Soviets," said Point Barrow trader Charlie Brower.

Clues to the loss of Levanevsky's plane include a report by Eskimos at Barter Island that they had heard an airplane flying easterly over Barter Island on August 13, 1937; another report came from Eskimo hunters who said they saw a big airplane splash into the sea near Oliktok Point on Alaska's arctic coast.

Expeditions attempting to find relics of the lost plane took place as recently as April, 1980, and in August and April, 1987. Earlier searches were made in the sea near Oliktok Point with a magnetometer.

From the North Pole all directions are south. Flying a course that is even a few degrees off can carry a plane far from its goal. A magnetic compass is unreliable near the Pole. In addition to a magnetic compass, the Russian plane had a sun compass (as well as possibly other navigational devices). The sun compass is useless when an airplane is in the clouds.

In 1947, during a weather data-collecting flight from Alaska's Eielson Air Force Base, near Fairbanks, to the North Pole and return, pilots of a modified World War II B-29 bomber, caught in weather, made a few degree error in navigation after they rounded the pole and flew south. Instead of arriving in Alaska, the plane landed in Greenland.

Could Levanevsky have made a similar slight error in navigation after he crossed the Pole and perhaps crashed in Greenland, or elsewhere?

The mystery of what happened to the N-209 has never been solved. However, Joe Crosson, Harold Gillam, and Sir Hubert Wilkins represented Alaskans well with their payback to the Russians.

Book Four

The Rest of the Story

The following chapters outline the aviation careers of pilots who, between 1923 and 1930, pioneered commercial flying mainly from Fairbanks, and who were involved in the Eielson/Borland search.

15
Joe Crosson

HE WAS ONE OF ALASKA's most famous bush pilots during the 1930s, nationally, and even internationally known. He was tall and handsome, with a personality smile. He was modest. When Washington D. C. politicians wanted to award him the Medal of Honor and the Distinguished Flying Cross for one of his flying exploits, he informed Alaska's delegate to Congress that he wouldn't accept either. "Not at all in keeping with what I did," he said.

The press sometimes referred to him as "The Mercy Pilot," because of his many medivac flights (a modern term for flying the medically challenged to a hospital). He disliked the appellation. "I'm just a pilot doing what I know best," he said.

Joe Crosson, born in Kansas in 1903, first saw an airplane when he was 10. It was a pusher (prop at the rear) at a local fairground, and he and his sister Marvel watched it take off and circle overhead. They became excited. "I'm going to be an aviator," he told Marvel, as he leaped about. She, three years older, also decided at that moment to become an aviatrix.

Their course was set, and it didn't waver.

Joe was an aviator like Joe Louis was a boxer, like Babe Ruth and Lou Gehrig were baseball players, and Jim Thorpe was an athlete. He became one of the best, and in time he rubbed shoulders with aviators Jimmy Doolittle, Eddie Rickenbacker, Billy Mitchell, and explorer Sir Hubert Wilkins.

The Crossons, parents Esler and Elizabeth, and Joe and Marvel, moved to San Diego, California. As young adults Joe worked as a mechanic in a garage, and Marvel clerked in a camera store. They saved their money to buy an airplane.

Joe Crosson with his typical happy expression. He's standing on a ski of Fairchild 71 NC9153 in April, 1930, when he was Chief Pilot for Alaskan Airways, Inc.

The Crosson Airplane

They bought for $150, and assembled, a stored former Navy N-9, the Navy version of the Jenny (JN4) trainer, powered with a Curtiss OX-5 engine for which they paid $125. While working on the plane, Joe took flying lessons. He soon got his license, and when their Jenny was completed, they towed it to San Diego's Dutch Flats Flying Field, where, in 1927, another pilot was to reach for the sky and fame with an airplane called the *Spirit of St. Louis.*

Joe taught Marvel how to fly. The year was 1922, and women pilots were looked upon as strange birds. With their own airplane, Joe and Marvel both built flying time, and within a few years both became skilled pilots. Joe learned aircraft maintenance by keeping their Jenny airworthy during those years. They kept their jobs and barnstormed with the Jenny to make money to pay for their continued flying.

Far to the north, in Fairbanks, tall, gangly and rough-on-the-edges A.A. Bennett was hired in 1925 by Jimmy Rodebaugh of the Fairbanks Airplane Corporation. He had, for fourteen months, been the manager of the San Diego Airport and Flying School. Rodebaugh had even taken flight instructions from Bennett. Bennett and Noel Wien were the two pilots for the Corporation. However, Wien left the company that fall, after deciding he didn't want to fly airplanes maintained by Bennett. He and his brother Ralph started their own company, flying from Nome the following year. Bennett then hired his San Diego friend Joe Crosson as a mechanic-pilot for the Fairbanks Airplane Corporation.

Flight to Teller

On May 15, 1926, *Pathe News* and the *New York Times* chartered a Fairbanks Airplane Corporation airplane to fly to Teller to get photos of the *Norge,* the dirigible of the Amundson-Ellsworth-Nobile expedition, which had just arrived there after flying over the North Pole from Kings Bay, Spitsbergen.

Joe Crosson made the flight.

A.A. Bennett, of the newly formed Bennett-Rodebaugh Company, (Rodebaugh and Bennett had left the Fairbanks Airplane Corporation to form their own company) was chartered by *International Newsreel* to fly to Teller, also for photos of the *Norge.*

Both planes successfully reached Teller, and both planes

departed Teller at the same time. Bennett's ninety mph Waco biplane arrived back at Fairbanks with film an hour before Crosson's eighty mph Standard biplane. Bennett then flew the film on to Whitehorse, where, by rail, it reached the sea and arrived in Seattle one day before the film that Crosson had flown.

The "race" between the two pilots to get the film Outside first for a scoop drew much interest in the news media; for example, it received an 18½-inch column on the front page of the May 17 *Fairbanks Daily News-Miner.*

Two months after arriving in Alaska, Joe Crosson flew to Teller and back with apparently no difficulty. The straight-line distance Fairbanks-to-Teller is about 560 miles. To get there his plane had to fly over a multitude of winding dog team trails; there were no roads, no railroads, nor signposts, nor other markings to show the way. There was an occasional small village. He flew down valleys, climbed over mountains, crossed treeless tundra, and vast spruce/birch forests. The land he flew over is essentially unchanged today.

Flights to Villages

During 1926, Crosson's passenger and freight flights spiderwebbed across mainland Alaska from Nome and Moses Point, both on the Seward Peninsula, to Wiseman, Bethel, McKinley Park, and, according to his logs, to more than twenty-five other named villages or mining locations. He flew his company's Standard, the re-assembled Jenny previously owned by Ben Eielson, and a new Super Swallow biplane that arrived in September that year. He also flew Waco 9s that belonged to the Bennett-Rodebaugh company (in the early years it wasn't unusual for a pilot to fly for competing companies; there weren't that many pilots, and business was business).

During that year he also walked more than 300 miles, returning to Fairbanks from broken airplanes left for repair in bush Alaska.

Waco in the Water

On July 24, 1926, while flying one of the Bennett-Rodebaugh Company's new Waco 9 biplanes, Crosson flew Joe Quigley to his mine in the Kantishna mining district near Denali (Mt. McKinley).

Pilot Ed Young went along for the ride. Quigley had assured Joe there was a "fine landing area," near the mine.

It wasn't a "fine landing area," unless you were talking about birds. It was a stump and rock-filled hump. Quigley probably figured if there weren't any trees, it could be used by a plane.

Crosson circled and found a gravel bar of the nearby Moose River that looked possible. He flew low and slow over it for a careful inspection. Setting up a landing, he reduced throttle and approached carefully, used some right rudder, and left and forward stick to slip and lose altitude. He poked his head out the left side of the plane, peering at the bar through his goggles.

It is impossible to see the ground straight ahead from a cockpit-equipped airplane; slipping it sideways allows a pilot to peer ahead to keep his landing spot in view until the last few moments.

Just before touching down, he straightened the plane with left rudder, centered and pulled the stick to his belly, and yanked the throttle to idle the engine to make a three-point touch-down. The wheels hit hard on pavement-like gravel. The airplane bounced, and headed for the river. Joe rammed the throttle wide open, hoping to climb back into the air, but the Waco stopped abruptly in the water anyway, perched at a 45-degree angle, nose-down and more upside-down than right-side-up.

Joe Quigley's nose was gashed and bleeding freely. The two pilots were unhurt. They climbed out and waded ashore. After a bit they pulled the airplane out of the water. Both propeller tips were bent, and the crankcase was busted. They were grounded and afoot 150 miles from Fairbanks. There was no phone, no radio.

Fannie Quigley, Joe's wife, poured iodine on her husband's raw sliced nose, and while the blood oozed, managed to sew him back together using ordinary thread. She had nothing with which to deaden the pain.

In all of Joe Crosson's years of flying, with around 8,000 hours of air time, Joe Quigley's torn nose was the only injury ever sustained by any of his passengers.

From the Kantishna Quigley mine, Crosson and Young hiked a nightmarish five days through eighty-five miles of brush, millions of mosquitoes, swamps and acres of tippy tussocks. They reached the Savage River camp at Mount McKinley National Park. From there they hitched a ride on the park road to the Alaska Railroad, which took them back to Fairbanks.

Crosson later flew mechanic George King to repair the broken Waco, this time landing on a six-mile-distant but much larger bar of Moose River. He flew the repaired Waco back to Fairbanks in mid-August.

The Super Swallow

Joe flew the company's new Super Swallow for the first time on September 23, 1926. The 180 hp Hisso-powered, two-cockpit biplane had three gas tanks that held ninety gallons, an oil radiator, adjustable radiator shutters and streamlined wires. It could take off within 100 feet and climb to 20,000 feet. It landed at a slow thirty-five mph. It was a well-named, modern (for 1926) ship, but still a cold ride for pilot and passengers.

It proved to be *really* cold when, on November 9, Joe took off from the river bar at McGrath with Iditarod miners Charlie Salmi and Carl Carlson and their end-of-the-season cleanup of $35,000 worth of gold. Joe got the Swallow airborne all right, but while the plane was still low over the Kuskokwim River, the Hisso quit, plunging the plane midriver into the water amidst chunks of drifting ice, which kept hitting the plane as they drifted downstream.

Fortunately, the plane had landed on a submerged sandbar. It squatted half submerged, ice floes streaming past. No one was hurt. Joe and the miners stood on the plane. They didn't dare leave it; the current would have swept them swiftly downstream to their deaths among the ten-to-twenty-feet-across chunks of foot-thick ice.

Residents of McGrath dragged a boat to the river, poled it to the plane, and ferried the three men ashore where they were dried and warmed by a huge fire. The rescuers retrieved the gold, and tied a line to the nearly new $4,000 airplane so it could be yanked ashore by a team of horses. It arrived ashore with the tail off and the wings heavily damaged.

The Super Swallow was shipped to Fairbanks in pieces. By March, 1927, it was once again airworthy. That Super Swallow and Joe Crosson were to be linked forever to one of the most remarkable northern Alaska flights of the 1920s. More on that later.

Fire Aloft

Crosson's next adventure, eight days later, could have easily ended his life. In October, 1926, he was flying the Fairbanks 1 Standard biplane, which A.A. Bennett had modified by building a canopy over the front cockpit, making room for four passengers. While he was aloft over the north slope of the Alaska Range, an engine's connecting rod broke, forcing Joe to land deadstick on a gravel bar of the Toklat River.

On November 13, he and mechanic Ernie Franzen were dropped off at the broken plane with parts and tools so they could repair it. Night time temperatures fell below zero and daytime temperatures climbed to a chilly ten degrees. They alternated between working on the airplane, and warming themselves by a fire. At night they pushed the fire away, and slept curled in their sleeping bags on the warmed ground. It required three days to make the Standard airworthy.

On November 15, with Franzen aboard, Joe flew the repaired Standard to the Diamond Roadhouse, where they spent the night. Next day they flew to Lake Minchumina, where they overnighted and luxuriated at another warm roadhouse. On November 17 they pointed the plane's nose toward McGrath.

Aloft at 1,500 feet near Medfra, thirty miles from McGrath, the plane caught fire. Joe pointed the nose at the nearest frozen lake. As the fire consumed the fuselage, Franzen held burlap between Joe and the fire. Crosson sideslipped the old Standard, to one of the fastest landings he ever made.

Once the plane stopped, both leaped out and stood helplessly on the ice and watched it burn. Now they were stranded 240 miles from Fairbanks.

They walked a few miles along a dog sled trail to the roadhouse at Berry's where they learned that pilots Ed Young and A.A. Bennett were both also grounded. The wings of Young's Waco were so damaged on a forced landing that he had left it, and started walking. He was given a ride by Bill Burke, the mail [dog] team driver and arrived at Fairbanks December 3. A.A. Bennett, landing on Berry's bar about fifty miles from McGrath, broke a propeller when a wheel broke through the ice and caused his Whirlwind Swallow to nose down. He was waiting for a dog team that was bringing him a new propeller. Thus all three pilots were afoot. This meant Joe and Franzen couldn't be flown to back Fairbanks.

A Long Winter Hike

To reach the railroad, Joe and Franzen had to walk 200 miles of dog sled trails to Nenana, the nearest train depot. No dog team traveled their way. It took eleven days. They spent some nights at roadhouses, other nights they built a fire and warmed the ground for a place to sleep.

On the trail, they came upon a big sign near a roadhouse, "Aviation Business Not Solicited," it read. At first they laughed. They didn't stop. Clearly, airplanes were affecting that roadhouse's business, as it was most other roadhouses in Alaska's Interior. Within a few years most roadhouses, and dog teams carrying mail, would disappear.

Rivers were frozen, snow was deepening, and at times the temperature dropped to -40 F. Fortunately, they were dressed in furs for flying in open cockpits.

It wasn't a fun hike. They were exhausted when the train, with them aboard, finally arrived in Fairbanks on December 7.

Crosson's Legendary Flight to Barrow

In 1927, Barrow was beyond the easy reach of most Fairbanks airplanes; besides being nearly 550 miles distant, it was on the far side of the huge, rugged, Brooks Range.

In 1926, Eielson and Wilkins had made three round trip flights from Fairbanks to Barrow with the single-engine Fokker *Alaskan;* Major Lanphier, with Wilkins, had flown the trimotor Fokker *Detroiter* to Barrow on one round trip. No other airplanes had ever flown the foreboding route.

Now, on March 25, 1927, at Fairbanks, Wilkins' new *Stinson Detroit News No. 1* and *No. 2* were loaded and ready for Ben Eielson and Alger Graham to fly them to Barrow. From there Wilkins could carry out his aerial explorations. But there was a problem. There wasn't room in either of the heavily loaded ships for A. M. Smith, the *Detroit News* correspondent. His newspaper was paying the bills, and the editors had sent Smith to write exclusive firsthand reports about the expedition.

Smith arranged with the Fairbanks Airplane Corporation for Joe Crosson to fly him to Barrow in company with the two Stinson Detroiters. Joe agreed to make the flight, for which he had three hours notice.

Joe flew the rebuilt Super Swallow C2375, with the liquid-cooled 180 hp Curtiss-Wright Model E Hisso engine. He had been flying the airplane for about a week since it had emerged from the shop after a total rebuild from its dunking in the Kuskokwim River. It carried enough gas for four hours in the air, not enough to reach Barrow. Joe would have to land at Wiseman to refuel. A welcome addition to the airplane was a heater in the pilot's cockpit, installed by master mechanic Jim Hutcheson, who figured out a way to circulate the engine's cooling liquid through a small radiator.

Joe had previously flown to Wiseman, so that wasn't a problem; however, he had never flown beyond there, across the towering peaks of the Brooks Range, and the hundreds of miles of the barren-appearing North Slope to coastal Barrow. There were no accurate maps. Joe, with Smith aboard, left Fairbanks two hours before the two *Stinson Detroiters*, planning to refuel at Wiseman. He would join the Stinsons as they passed overhead bound for Barrow.

It didn't work that way.

Joe landed at Wiseman and refueled, as planned. When he restarted the engine, the radiator had apparently frozen, splitting some seams. Anyway, it now leaked.

When the two *Stinson Detroiters* circled overhead expecting him to join them, Joe waved them on. He drained oil from the Swallow's engine and kept it warm in the roadhouse. He then removed and soldered the radiator so it no longer leaked.

Off again the next morning with his passenger, Joe flew at 8,000 feet for more than an hour to cross the high peaks of the Brooks Range. A dense haze lay over the North Slope, and the temperature dropped to -40 F. His magnetic compass pointed to the magnetic pole, just north of Canada, roughly thirty degrees from true north. He judged his direction of flight from the sun as he flew northwest, following oral directions given him by Ben Eielson.

He arrived at the Arctic Ocean about fifty miles east of Barrow, followed the coastline west, and landed at Barrow four and a half hours after leaving Wiseman.

His feat impressed Wilkins, a fact which became significant later in Joe's career when Wilkins selected pilots to accompany him to Antarctica. Wilkins had assumed Joe had returned to Fairbanks from Wiseman, and was amazed and pleased that, with reporter A. M. Smith, he had managed to safely reach Barrow on his own.

The Return Flight to Fairbanks

Crosson remained overnight in Barrow, and started his return to Fairbanks February 28. Deciding not to cross the Brooks Range again, he flew southwest, following the coastline. Near Cape Thompson, the radiator started to boil. He landed on a stretch of smooth ice, left the engine running, and chipped a quantity of ice and fed it into the radiator, working within inches of the spinning prop.

That cooled the engine. He took off and flew on. After nearly four hours, he landed and refueled with case gas he carried in the front cockpit. Six hours after leaving Barrow, he arrived at Kotzebue where he landed, to the excitement of the Eskimos, some of whom had never seen an airplane. A blizzard held him there four days.

While he was at Kotzebue, a Native from Noorvik arrived. His mission was to send a radio message to Fairbanks requesting an airplane be sent to Noorvik to fly out the wife of Dr. Ventner, a government physician stationed at that village. Thus Joe could pick up a paying passenger on his way home.

When Crosson landed on the rough river ice at Noorvik, his landing gear broke. His was the first airplane to land at this Eskimo village located on the clear Kobuk River forty miles inland from Kotzebue. The local Eskimos, experts at repairing almost anything, helped him fix the landing gear.

With Mrs. John Ventner as a passenger, he left Noorvik the next morning, taking off into a strong surface wind. The Super Swallow lived up to its name by leaping into the air within about fifty feet. Once aloft, he climbed to cross the low, cloud-covered, Waring Mountains. A strong tailwind pushed him to Ruby in less than two hours, where he refueled, and went on to Fairbanks with Mrs. Ventner. For the last fifty miles, from Dunbar, a station on the Alaska Railroad, due to poor visibility he flew a few feet above the tracks.

Without a map, in eight days, in March, when the Arctic was still in the throes of winter, he had flown an open cockpit airplane a circle of 1,580 miles over mostly uninhabited wilderness. Along the way he had repaired his radiator and landing gear. It was an almost incredible achievement that has been hailed as one of the outstanding flights in the Alaska of the 1920s.

Marvel Arrives in Alaska

Marvel Crosson arrived at Fairbanks in October, 1927. She soon became acquainted with pilots and mechanics. She was given a check ride by Department of Commerce check pilot Ben Eielson and earned a limited commercial pilot's license, making her the first woman licensed to fly in Alaska. She could now legally carry passengers.

She started flying with some of the pilots, and logged nearly fifty hours in the Alaskan air. She flew mostly short hops, but she did make flights to Anchorage and Nome. Her 1927-28 flights were the first ever by a woman in Alaska.

Joe Goes East

While Marvel was working her way into the flying business at Fairbanks, Joe made a trip to the East Coast. He and Ben Eielson had discussed the need to find capital to form a large and more efficient flying company in Alaska. He sought investors in New York, but found they were elusive. He ran out of money and got a job as a pilot flying Fokker Universal airplanes on floats in Ontario for Western Canada Airways. He quickly became a competent float plane pilot.

To Antarctica

After his dramatic Barrow to Spitzbergan's flight, Wilkins had no difficulty in finding backers for his plan to make aerial exploration flights in Antarctica. He chose Ben Eielson and Joe Crosson as his pilots, and aviation mechanic Orville Porter to care for the two

Lockheed Vegas he planned to take. Porter, a member of his 1927 expedition with the Stinson Detroiters, had remained in Alaska.

In August, Joe traveled to Los Angeles to check Wilkins' new Lockheed Vega, named *San Francisco*, and to fly it to New York. The Barrow-to- Spitzbergan Vega X3903, now named *Los Angeles,* was already at New York, ready to be loaded aboard the Munson Liner *Southern Cross,* the ship that was to take the expedition to Montevideo, where they would transfer to the whaling mother ship *Hectoria*, which was to take them to Antarctica.

In Antarctica Eielson flew the main exploratory flight with Wilkins from Deception Island. Afterward, on January 10, 1929, Joe flew Wilkins on a 250-mile exploration flight, searching fruitlessly for a better airplane landing site.

Joe Crosson was the second pilot ever to fly an airplane in Antarctica.

Ferry Pilot

Upon return to the U.S., Joe worked at ferrying Lockheed Vegas for the Los Angeles factory to about a dozen widely scattered states from New York to New Mexico. The new and advanced design of the Vega had taken the world of aviation by storm. It was fast, efficient, and beautiful, and many a pilot made a name flying Vegas on long distance flights, as well as in major races. Today, more than three quarters of a century later, the airplane is still a standout.

On one of his ferry flights Joe carried a hitchhiking pilot named Wiley Post, who was to make his Vega, the *Winnie Mae*, one of the most famous ever. He twice circumnavigated the globe with it. Joe Crosson and Wiley Post became close friends.

Marvel's Triumphs

In California, Marvel won the first National Aeronautics Association Women's Air Race, flown from Palo Alto, across San Francisco Bay, to Oakland. She also set a new woman's altitude record when she reached 23,996 feet flying a Ryan Brougham cabin monoplane.

In August, 1929, twenty of the world's best women pilots competed in the first Women's Air Derby from Santa Monica, California, to Cleveland, Ohio, with nine designated re-fueling and rest stops. Marvel Crosson was favored to win, for she flew a clipped wing Travel Air provided by Walter Beech, one of the partners in the Wichita, Kansas Travel Air factory. It was the fastest entry and had been clocked at 168 mph.

On August 19 she departed Yuma bound for Phoenix, but she didn't arrive. Next day her crashed Travel Air and Marvel's body was found, her parachute beneath her. She had apparently bailed out and pulled the ripcord, but the chute hadn't fully opened.

Joe was crushed. His life-long flying partner was gone.

Alaskan Airways

Joe returned to Alaska as chief pilot of Alaskan Airways, the new company that Ben Eielson had put together. Eielson was the manager. It was a leap forward for Alaska's fledgling aviation industry.

An opportunity for the new Alaskan Airways arose that November of 1929, when Siberian fur trader Olaf Swenson contacted Fairbanks and arranged to have fur flown to Alaska from the Siberian coast and his ice-bound trading ship *Nanuk.*

When Ben Eielson and Earl Borland left Teller on November 9, 1929, Joe Crosson was busily flying out of Fairbanks. By the end of November when there was no word on the missing Hamilton, Joe knew he had to start a search. At the time Noel Wien, Ben Eielson, and Joe Crosson were the only pilots in Alaska who had extensive flying experience under arctic winter conditions. Wien was still in the states, where he had gone after selling his company to Alaskan Airways.

At Fairbanks, Joe said goodbye to Lillian Osborne,[1] who would become his wife in August of that year, telling her he wouldn't be back until he found Ben. He arrived in Teller in the open cockpit Waco 10 biplane on November 29. As related earlier, it was December 19 before the weather allowed Joe and Harold Gillam to fly to Siberia.

1. Author: I was privileged, with my family, to entertain Lillian Crosson, Joe's widow, for an evening in my home at College, Alaska, in 1953. She was a lovely, charming, woman.

Crosson was still mourning the loss of Marvel. The loss of Ben Eielson was another blow. He and Eielson had become close friends while with Sir Hubert Wilkins on their Antarctic venture. Earl Borland, too, had been a good friend.

Wings Over Alaska

Airplanes fit Alaska. By 1936 there were seventy-nine airplanes flying commercially in the Territory – a big jump from the twenty-four planes found there in 1930. Joe Crosson, married in 1930, settled down and became one of the Interior's most popular and most active bush pilots.

He achieved many firsts, and made many flights that brought him praise.

On March 7, 1931, he made another flight to Barrow from Fairbanks, flying the Wien open-cockpit Stearman biplane to deliver diptheria serum. An epidemic had started in Point Hope at Christmas. Two people had died from the disease at Wainwright. One case had recently developed in Barrow.

This time he had an air-cooled engine which was far more reliable than the liquid-cooled Hisso that took him to Barrow in 1927. His flight over the Brooks Range and Arctic Slope went without a hitch.

More serum was needed, and, flying through a storm, he delivered another batch on March 13, this time flying a Fairchild 71 cabin plane.

Wiley Post

In 1933, when Wiley Post made his second flight around the world, solo this time, he became lost after arriving in Alaska, and landed on the 700-foot-long runway at the tiny mining town of Flat. He ran off the end of the runway into a ditch, bending his prop and breaking his landing gear.

A radio message to Joe Crosson at Fairbanks brought Joe and two mechanics in a Fairchild 71 with a prop and tools to repair his friend Wiley's hot, all white *Winnie Mae*.

As a result of Joe's help, Wiley Post set a new world record for circling the globe. Afterward, Wiley returned to Alaska, and he and Joe hunted and fished together. They had been friends since Wiley had hitchhiked a ride with Joe in 1929 when Joe was delivering Lockheed Vega airplanes for the factory.

Mercy Pilot

Although Joe Crosson didn't care for the "Mercy Pilot" title the press used when referring to him, during his years of flying from Fairbanks he probably flew more medivacs than most bush pilots, for Fairbanks, with doctors and a hospital, was the hub of dozens of villages, mines, and bush residents. When anyone in the surrounding bush urgently needed medical care, they went to Fairbanks, and Joe often flew them there.

An example of Crosson's response to a medical emergency occurred in the fall of 1934 when Theodore Van Bibber became seriously ill at his cabin on Wood River, about ninety miles south of Fairbanks.

Fairbanksans Bob Buzby, Keith Harkness, Roy Madson, and Bill Cory, were on a November mountain sheep hunt in the Wood River area with two dog teams. They found Van Bibber at his Wood River cabin in agony and having seizures. Van, a powerful six-foot-two, was a well-known resident of Fairbanks who, for many years, ran a dog livery at his Rabbit Island Fairbanks home caring for teams of sled dogs of bush residents while they visited and shopped in town.

The hunters suspected he had appendicitis, and put cold packs on his stomach. After observing Van Bibber for a few hours, Bob Buzby decided to take a dog team to Fairbanks and arrange for an airplane to pick him up. He feared the ill man might not survive the trip to Fairbanks riding in a bounding and jostling dog sled.

Buzby and Harkness, with fifteen dogs, struggled for thirty hours, fighting overflow on Wood River, and deep snow drifts, and dropping temperatures to reach Fairbanks, arriving there when the temperature was -55 F.

Joe Crosson agreed to fly to Wood River to try to pick up Van Bibber with a warm Fairchild 71 he had in a hangar. Long-time Fairbanks physician Dr. Gillespie agreed to accompany him on the flight, and Bob Buzby rode along to guide Joe to Van Bibber's cabin.

Daylight hours were short, and they hurriedly warmed oil and poured it into the Fairchild. The airplane swiftly reached Wood River. Joe landed in a precarious place only because he knew Van Bibber was counting on him. With the airplane stopped, Buzby put two-by-fours in front of the skis so Joe could taxi up on them. That kept the skis from freezing down.

While Buzby and Dr. Gillespie hurried to Van Bibber's cabin, Crosson put the canvas cover over the engine, drained the oil, and lit the plumbers' fire pot to keep the engine warm, and occasionally put the oil on it to keep it warm too.

With Van Bibber aboard, Joe poured the warm oil back into the big Wasp engine and got it started. He had to make his takeoff run downstream. It was close, but he made it and reached Fairbanks in twenty-five minutes. It was dark when he arrived, but a ground crew put out railroad flares while Crosson circled. He landed safely.

Van Bibber recovered from a strangulated hernia at Fairbanks's St. Joseph's Hospital.

A team of fifteen dogs and two strong men had struggled for thirty hours to travel the distance the Fairchild 71 covered in twenty-five minutes. Despite the fifty below zero temperature, Joe Crosson met the challenge and probably saved Van Bibber's life.

Although Crosson didn't care for it, perhaps "Mercy Pilot" was a fair description of him for his many medivac flights. No doubt Theodore Van Bibber would have thought so.

Wiley Post and Will Rogers

In August, 1935, Wiley Post flew his low wing, overpowered, nose- heavy float-equipped airplane to Fairbanks. His companion was nationally popular humorist, movie star, and columnist Will Rogers. Joe flew them around the area for a couple days, showing them Alaska. On their last night at Fairbanks Lillian Crosson cooked dinner for them, and on August 15 Wiley pointed the nose of the big plane north as they flew toward Barrow.

On the arctic coast they ran into fog, and Wiley wasn't sure where they were. He landed on the Walapka River to ask directions of Eskimo Claire Okpeaha who was camped nearby. Okpeaha pointed toward Barrow, a mere twenty miles away, and Wiley took off.

(l. to r.) Will Rogers, famed dog musher Leonard Seppala, Wiley Post, and Joe Crosson, August 14, 1935, next to the Wiley Post plane on the Chena River at Fairbanks. Rogers and Post were killed the following day when their plane crashed near Barrow.

At about fifty feet the big engine stopped, the plane stalled, and crashed nose down, killing both Post and Rogers.

Claire Okpeaha waded to the plane and found no life. He then ran the twenty miles to Barrow with the news, which was sent by the Barrow radio to the world.

When Joe Crosson learned of the deaths, he asked Bob Gleason to accompany him as radio officer on the flight to Barrow to claim the bodies (Gleason, the one-time radioman of the *Nanuk,* was now organizing aviation radio communication in Alaska). They loaded a float-equipped Fairchild 71 with extra case gas so they could refuel on the way, and headed north. As they flew, a wire from Charles Lindbergh arrived at Fairbanks, asking Joe to fly to Barrow to get the bodies.

The bodies had been taken to Barrow by boat. Joe loaded them aboard the Fairchild at the Barrow lagoon, and added six cases of gas (sixty gallons) to the load. He fought bad weather through the Brooks Range, and landed at Wild Lake on the south slope of the range to add the sixty gallons of case gas.

From Fairbanks Joe, with copilot Bill Knox, flew a Lockheed Electra twin engine plane to Seattle with the bodies. Bob Gleason

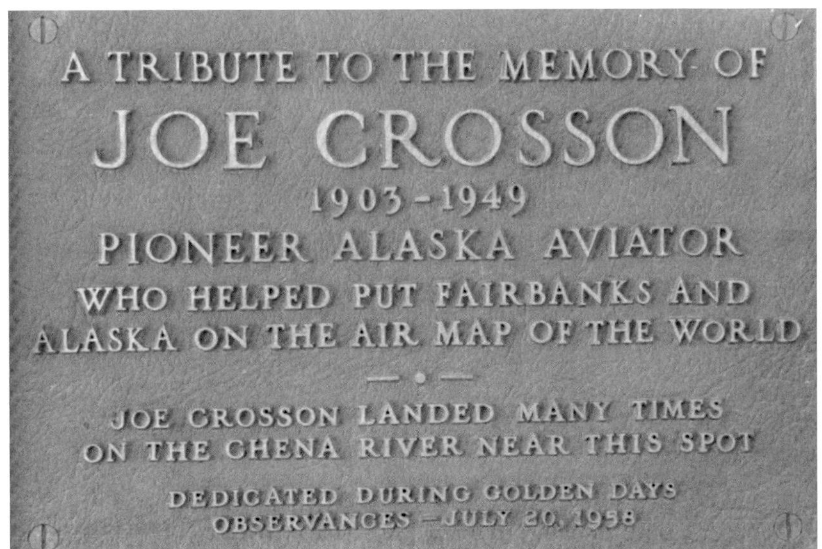

This plaque memorializing Joe Crosson may be seen at Fairbanks on the bank of the Chena River, viewable from First Avenue, near the Cushman Street bridge. AUTHOR

flew as the radio flight officer. At Seattle the bodies were transferred to a Pan American DC-2 which flew to Burbank, California, where Will Rogers' body was left; the plane then flew Wiley Post to Oklahoma where he was buried.

The loss of his close friend Wiley Post, coming so soon after the deaths of Marvel and Ben Eielson, had a great impact on Joe, although he handled his grief by himself. It was a private matter he couldn't share.

To Joe, it was almost an insult when members of Congress wanted to award him medals for flying the bodies of his friends from Barrow to Seattle. He respectfully declined the honor.

Pacific Alaska Airways

Pacific Alaska Airways, a branch of Pan American Airways, was Joe's baby. It became the first modern airline in the Territory. The airplanes were equipped with radios, as were the various stations. A big new hangar was built at Fairbanks. Scheduled flights were established, including service between Alaska and the states. Joe both flew and administered, which kept him busy through the late 1930s.

By 1941 mail and passengers could travel from Seattle to Fairbanks in one day. Joe and his family now lived in the Seattle area where he was named the Division Manager for Pan American Airways of Pacific Air Alaska. When he started having health problems, he resigned in early 1944.

He acted as an advisor to the military during the Aleutian Campaign of World War II. With several partners, including Noel Wien (then retired from flying and living in Seattle), he bought Northwest Air Service, a company that provided parts and maintenance for aircraft, with a hangar at Boeing Field, Seattle.

Joe Crosson died of a massive heart attack at the age of 46 on June 21, 1949.

In Anchorage, Crosson Circle and Crosson Drive lie a few blocks north of Lake Spenard, and near Borland Drive, Will Rogers Place, Wiley Post Loop, and Gillam Circle.

In 2002, Joe Crosson was inducted into the Alaska Aviation Heritage Museum Hall of Fame.

At Fairbanks, a bronze plaque on First Avenue, near the Cushman Street Chena River bridge, memorializes Crosson's contributions to Alaska.

16
Harold Gillam

The Legend

"I don't know how in hell he did it!" was a common refrain among early bush pilots of Alaska's Interior when commenting on the flying exploits of Harold Gillam. He seemed impervious to bad weather. Repeatedly, he arrived safely after boring through clouds, fog or snow storms that kept other pilots grounded.

An oft-repeated story about Gillam has a handful of glum bush pilots grounded by a blinding snowstorm at McGrath, one of Gillam's 1930s Kuskokwim Valley mail route stopping points. They were tired of poker, tired of the roadhouse, tired of the storm and tired of each other's worn-out stories. All had read and re-read the roadhouses' pile of last year's magazines. Their wind and snow-blasted airplanes were tied down outside the roadhouse, almost invisible through heavily falling snow. They had waited out the storm for days.

In the roadhouse's main gathering room they stopped what they were doing to listen to the unmistakable, high-pitched whistle of an American Pilgrim – its prop-tip speed was much higher than that of other bush planes. The pilots crowded around a window, each knowing precisely what they would see. As they watched, Gillam's big, silver, pot-bellied ski-equipped American Pilgrim appeared from the thickly falling snow and neatly touched down and taxied to the roadhouse. Gillam, in furs, and looking like an arctic explorer, climbed out of the plane, unloaded mail sacks and packed them into the roadhouse.

Harold Gillam in the cockpit of the Wien Stearman NC5415 which he flew to Siberia, and used in the search for the lost Hamilton.

An American Pilgrim, identical to the three Pilgrims once owned by Harold Gillam. First built in 1931, at the time the Pilgrim was one of the largest airplanes made in the United States. This Pilgrim, N709Y, is now one of the treasured airplanes at the Alaska Aviation Heritage Museum in Anchorage where it is being restored. It was flying hunters into the hills, and flying back with moose meat when this photo was taken of it at Yakutat, Alaska, circa 1965. Ed Huizer

"Gas her up," he told the local roadhouse handyman. "I'm going on to Fairbanks."

While waiting for his plane to be re-fueled, he drank a cup of coffee, stood briefly near the wood-filled Yukon stove, and watched the poker game.

When the big airplane was ready, he left. He hadn't said a word to any of the other pilots. Gillam was not talkative.

The poker players paused when they heard the roar of the Pilgrim taking off and flying into the storm.

The roadhouse radio, tuned to an aircraft frequency, didn't make a sound for the next several hours. Snow still filled the air next morning. Still, there was nothing from the radio but weather reports. Did Gillam make it to Fairbanks in the terrible storm? The pilots were concerned. Finally, they queried Fairbanks only to learn he had arrived on schedule. Gillam had been so unconcerned on his way home that he hadn't bothered to use his airplane's radio.

Some said that Gillam had iron nerves. Others claimed he had no nerves at all. "Gillam is part bird," others commented.

The story may be apocryphal. However many awe-struck pilots told tales of Gillam's bad weather flights. Undoubtedly he commonly flew in weather that other pilots refused to challenge. The report of Dan Cathcart, the United Airlines pilot who flew with Gillam from Fairbanks to Barrow, previously mentioned (chapter 14) is typical.

The Man

Gillam was medium-sized, muscular, dark-eyed and dark-haired, handsome. He was attractive to women. His movements were smooth, graceful. He had been an amateur boxer. He radiated power, but at the same time he seemed relaxed. He was proud of his appearance, and dressed neatly, and he was always clean. He somehow managed to remain clean even when working on his airplanes. He liked his airplanes to be clean, in contrast to some other early bush pilots.

Charles Harold Gillam was born in Kankakee, Illinois, April 8, 1902, and grew up in Chardron, Nebraska. At age 16 he ran away from home and joined the United States Navy. During his four year hitch he qualified as a hard hat deep-sea diver. He was discharged in California in the spring of 1923, and traveled to Seattle to visit his parents, who were now living there. While in Seattle, he signed up to work for the Alaska Road Commission for $150 a month. He became a cat-skinner for the ARC working on the Richardson Highway; later he drove tractors for the Fairbanks Exploration Company, a major gold mining company.

After that, Gillam bought his own tractors, and in winter dragged freight with them across the frozen land from Fairbanks to various mines.

From Tractors to Airplanes

With his tractors he helped build the local airport at Fairbanks. As he watched biplanes gracefully circling and landing, he became interested in flying. He went to San Diego and bought a Jenny biplane and shipped it to Fairbanks. He replaced the OX-5 engine with a more powerful and heavier Hisso, and started taking flying lessons in it from former Navy pilot Marcel L. "Danny" Danforth.

Gillam became the survivor of Alaska's first fatal airplane crash. Instructor Danforth was killed when Gillam's nose-heavy Jenny stalled and spun to the ground when the engine quit. Danforth died a day later. Gillam's back was hurt, and bones in his hands were broken, but he was soon flying again. He was one of the first half dozen students to solo and receive a pilot's license in Alaska.

Accounts vary on whether he had a pilot's license, and how much time he had logged in the air when he joined the search for Ben Eielson and Earl Borland (chapter 12). Some accounts claim he had no pilot's license. Others report that he received a limited commercial pilot's certification for single engine aircraft on April 11, 1929. Some accounts report his having logged forty hours, others report fifty hours, when he joined the search. Regardless, forty or fifty flight hours is minimal experience for a pilot, even today with safer, modern, easier-to-fly airplanes.

Gillam's life change came with his participation in the search for Eielson and Borland. His amazing flight through fog and clouds for 300 miles to the *Nanuk*, when Joe Crosson was forced back, and Crosson's generous praise for the feat, became well known across Alaska. The fact that he had flown on the search for the lost Hamilton impressed investors when, soon after returning from the search for it, he moved to Copper Center to pioneer commercial flying in the region.

Copper Center is not an easy place from which to fly. Storms constantly batter the area. Fog, clouds and mountain air turbulence are a constant challenge. Gillam started with a Swallow biplane, a Zenith six-passenger cabin plane with a cockpit for the pilot, and an Ireland Neptune amphibian. In 1931 he had six wrecks within six months. He spent so much on repairs that he was constantly broke. His backers lost more than $30,000 – big money at the time.

Gillam's Medivacs

Gillam seemed to like flying at night. The trader at Copper Center, John McCrary, fell ill and took to his bed, saying he was

Harold Gillam's Zenith biplane Z6-B, NC977Y photographed by Noel Wien at Copper Center, Alaska, in 1931. The Zenith company built farm implements in Midway, California. Alaska pilot A. A. Bennett convinced them to design and build this rugged biplane, starting in 1929. Only seven of this model of the Zenith were ever built (Gillam's was probably number 5). The depression halted production, and the company went back to making farming implements.

The cabin of this Zenith held four passengers, and Gillam flew it from the open cockpit. He hauled high grade ore from back-country mines with it, as well as the usual run of Alaskan freight and passengers. It was powered with a 420 Pratt & Whitney Wasp.

One other Zenith was owned and flown in Alaska by the Bennett Rodebaugh company at Fairbanks.

Gillam's Zenith met its end in 1936. Details vary on whether it was damaged while flying, or from a storm while on the ground.

fine. No one believed him. Word got to Gillam, who, despite a snowstorm, loaded McCrary into his Zenith and flew 120 miles to McCarthy and landed after dark on the short strip. McCarthy was five miles from Kennecott and a doctor.

"He's pretty sick," the doctor said, after operating and finding a perforated ulcer. McCrary had lost much blood.

Harold then flew through the snowstorm the 100 miles to Cordova to fly McCreary's son Nels, Cordova's fire chief, to his father. He landed the Zenith on the ice of Eyak Lake, refueled, loaded Nels, and, flew back through the snow storm to McCarthy, landing after dark.

McCrary survived. Harold Gillam had saved the father's life, and seen to it that the son reached the critically ill man.

In a letter to me (author) in December, 1979, Nicholas Lincoln, Sr., of Kenny Lake remembered that in 1929 he first saw Gillam when he and A.A. Bennett flew the Waco 10 NC780E biplane from Fairbanks and landed at Copper Center to refuel. The two had arrived to fly injured miner Jack Carroll, hurt on a cave-in at Chisana, to the Kennicott hospital. The date was October 3. Gillam was along as a guide for Bennett; he knew the region, having worked there three years for the Alaska Road Commission.

"Just about all the citizens of Copper Center gathered to see that 'flying machine,' as they used to call it," Lincoln wrote.

"My dad worked for Harold, and I had my first ride in his Swallow biplane on the 4th of July, 1934. What a thrill!

"When he used to make night flights from Fairbanks, Valdez, or Cordova, I often listened to him on my shortwave radio, as he gave his half hour position reports. I especially remember the night he made a mercy flight from Cordova to Nabesna. He left Cordova in late afternoon and flew his Zenith into a storm. He arrived at the Nabesna mine in the Wrangell Mountains and picked up owner Carl Whitham, who had fallen off an ore bunker and was seriously injured and needed immediate surgical attention.

"With Whitham aboard, he flew back through the storm and arrived at Cordova late that dark night to land safely on the narrow hillside strip that served as Cordova's airport.

"I remember the night he flew to Copper in pitch darkness and landed to the far left in the brush and stumps and nosed his Zenith over. Another time he ran out of gas just across the Copper River from Copper Center and landed in the brush with a load of ore from the Nabesna mine.

"He borrowed a boat and rowed across the river. While others were headed across the river looking for him, I helped him beach the boat. He walked back to his parked Ireland Amphibian, cranked it up, and flew over his rescuers back to the downed plane.

"I remember the pilots and mechanics who worked for him. Mostly I remember Oscar Winchell, the most lively and funny man of the group. There were also pilots Elbert Parmenter, Don

Cathcart, Bert Lien, Leo Moore, M.D. Kirkpatrick, and Johnny Moore; and mechanics Tom Appleton and Earl Woods."

I Don't Know How the Hell he Did it

Long-time Alaska bush pilot Rudy Billberg knew Oscar Winchell who flew for Gillam for a time. Winchell related to Billberg an experience he once had while flying above clouds with Gillam; "There were mountains below, with no breaks in the clouds. Gillam circled, and pointed the nose of the airplane down through the dense clouds. When we broke into the clear, we were directly over the airport.

"After Gillam landed, there wasn't enough gas in the airplane to taxi it to the hangar. I don't know how the hell he did it," Winchell said.

Gillam Airways

From his bases at Copper Center and Cordova, Gillam pioneered scheduled flights between Cordova, Fairbanks, and Valdez. He initiated countless out-of-the-way runs, bringing air service to many isolated villages and mines.

On June 2, 1930, Gillam, at Fairbanks, was given an airman's examination by Department of Commerce inspector Wylie R. Right. He passed and received his Air Transport Pilot License, so he was now legal to fly passengers, and could continue what he had been doing all along.

At Fairbanks: the Pilgrims

After about six years of flying in the Copper Center-Cordova region, Gillam was nearly broke. He borrowed money to buy a second-hand Hornet-powered American Pilgrim, and moved to Fairbanks. He suddenly stopped breaking airplanes until the 1943 crash that resulted in his death. Perhaps it was the fine flying quality of the Pilgrims (he eventually owned three of these great airplanes) or their loving care by Tom Appleton, his longtime mechanic. In the six years he worked for Gillam, there was not one engine failure with a Pilgrim.

The American Pilgrim, first built in 1931, was a slab-sided, pot-bellied appearing ship. It was built in the Fairchild factory by American Airplane and Engine Corporation, and was designed for one pilot – no co-pilot. Some Pilgrims carried ten passengers, others carried twelve. The ball-bearing-mounted controls were especially smooth. The airplane had a long thick wing and a long propeller. The tips of its prop traveled at or near the speed of sound and made a terrible racket; the sound of an aloft Pilgrim was unmistakable. That big airplane, one of the largest made in the U.S. at the time, leaped off the ground quickly and handled like a high-powered light plane.

It was a slow-flying airplane that could handle heavy wing icing. The pilot sat high in the front with wonderful visibility. It was very stable in the air and landed easily. While many early pilots were content with a compass alone as a flight instrument, Gillam

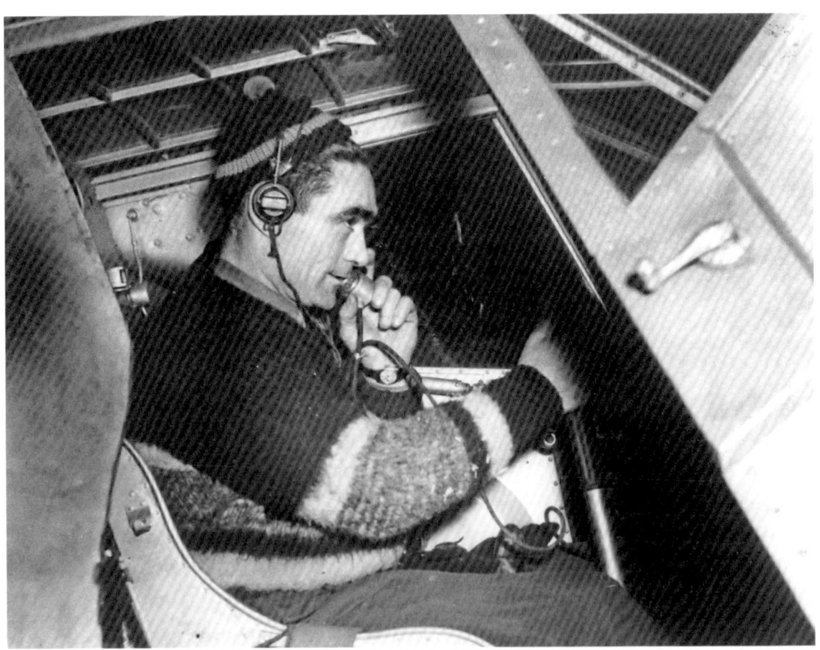

Harold Gillam in 1936, in the pilot's seat of one of his nine-place Pilgrams.

installed a directional gyro, a sensitive altimeter and an artificial horizon in each of his Pilgrims.

Pilots joked that the Pilgrim took off, flew, and landed at the same speed. Actually it cruised at 110 mph, and when empty landed at fifty or sixty mph.

His Pilgrims flew passengers between Cordova and Fairbanks. Most were Alaska Steamship passengers; from the steamers they commonly landed at, and departed from, Cordova. Many lived at or near Fairbanks.

Weather Bureau Contract

Gillam contracted with the U. S. Weather Bureau to make weather flights over Fairbanks. Between September 15, 1936 and March 15, 1937, he made twice-daily flights. The program was continued in 1937-38 with flights every third day. Each flight was to 16,500 feet. The rate of climb and descent had to be exactly 300 feet a minute. When the designated altitude was reached, Gillam descended in a tight spiral.

For these flights his Pilgrim carried an aerometeorograph, which recorded pressure, humidity, and temperature. When he landed, weathermen W. B. Drawbaugh and L. A. Coffin reported the readings to weather forecasters in the States. The reports were meant to give mid-westerners warning of weather changes.

Many of these flights were in complete darkness. At first, during these flights, mechanic Tom Appleton stood listening for the noisy Pilgrim, and with one of the first voice radios installed in a plane in Alaska, he told Gillam where he was in relation to the airport runway. This wasn't very reliable. Gillam, who always sought perfection, then installed at his hangar a battery charger that functioned as a short-range homing signal. With a simple direction finder in his plane tuned to the frequency of the noisy charger, he could determine the direction to the airport. It was the first such device used for aircraft navigation in Alaska.

Substitute pilot Bert Lien did the honors when Harold was unavailable. Only twice during the two years of contract were the flights cancelled, both times due to the condition of the airport runway.

The Kuskokwim Mail Route

In 1938 Gillam won a contract to fly a Kuskowkim Star Route winter mail run. This called for twenty-six weekly round trips between Fairbanks and Bethel, with eleven intermediate stops each direction for 525 miles. Pan American Airways had previously flown this route. Dog-team-carried-mail was more reliable than Pan Am, although slower; Pan Am pilots didn't fly in bad weather.

Gillam decided to deliver and pick up mail on a schedule. Once set, his schedule didn't change.

"We could set our clocks by Gillam's arrival," Don Harris, a long-time resident and miner at McGrath once told me (author).

He established a 100-percent completion schedule during the three years he flew this route, regardless of weather. He commonly flew it when other pilots remained grounded by weather.

Starting in 1949, Rudy Billberg, a pilot for Northern Consolidated Airlines, flew the Fairbanks-McGrath route hundreds of times over many years, with both a Douglas DC-3, and a T-50 twin-engine Cessna (the latter often, insultingly called a "bamboo bomber").

"I know what the weather can be like on that route," he once told me (author). "I don't understand how Gillam managed to establish such an unblemished record. McGrath residents told me that sometimes the old silver Pilgrim that Gillam flew came sneaking in just over the trees. Once, when weather pushed him down to the ground, he taxied the Pilgrim on skis ten or fifteen miles on the snow of the Kuskokwim River to reach McGrath," Billberg remembered. He also remembered that residents of the Kuskokwim valley adored Gillam.

According to Billberg, Gillam told Ray Peterson, a long time Alaska bush pilot and later President of Northern Consolidated Airlines, "If you can't see because of bad weather, go on instruments at 200 feet. Eventually you'll see a reference point."

Billberg was aghast at this statement. Barometric pressure changes, sometimes quickly. Though an altimeter might indicate a plane is flying at 200 feet, it might actually be at 100 feet or lower. Land elevation varies; even small hills can easily poke up 250 feet. At 110 miles per hour, the speed of Gillam's Pilgrims, the sudden appearance through his windshield of a nearby tree, a hill, or a mountain would give him virtually no time to react.

Gillam once told Billberg that while flying he could hear

echoes from the sound of his airplane coming from nearby hills or mountains. Billberg flew one of Gillam's Pilgrims through the summer of 1942 on the Northway operation (see following), so he was familiar with the high-pitched scream of the Pilgrim's prop, and the deep roar of the great 575-horsepower Hornet engine. He had difficulty believing Gillam's ability to hear echoes of the plane's noise while aloft in a Pilgrim.

The Northway Operation

The Northway operation involved construction of one emergency landing, waystop and refueling airstrip of the seven[1] that, starting in 1942, were built in Alaska during World War II. They were designed to aid the nearly 8,000 wartime lend lease fighters, bombers, transports and training planes flown from the States through Alaska to Russia to shore up Russian defenses against invading Germans.

When the Morrison-Knudson Company started to build these runways, they needed a pilot to fly workers, engineers, engine parts, plans – whatever - swiftly and safely between the various runway jobs. They named Harold Gillam as their Chief Pilot. In this position he flew a twin-engine Lockheed Electra. As he had with his Pilgrims and other ships, he unhesitatingly flew overloads. The Civil Air Administration, recognizing the war emergency and the need to swiftly prepare Alaska for whatever might be coming, largely ignored the overloads.

Gillam's Luck Runs Out

It wasn't an overload that ended Harold Gillam's career.

On January 5, 1943, Gillam, flying the all metal, twin engine M-K Lockheed Electra with five passengers, left Seattle bound for Ketchikan and Anchorage. His airplane was the same type flown by Amelia Earhart on her doomed last flight. A violent storm had slammed into Southeastern Alaska, and other Alaska-bound flights had been canceled because of it. Boeing Field clearance officers were reluctant to allow him to leave, but Gillam left anyway. He had earned an instrument rating at Houston, Texas, during the

1. Built at Northway, Tanacross, Big Delta, Tanana, Galena, Moses Point, and Nome.

summer of 1941.

Daylight was mostly gone at five that afternoon when the Electra was overdue and unreported at Ketchikan. Gillam had called flight controllers to report he was in trouble; one of his two engines had quit. He had entered fog four hours out of Seattle, and had gone on instruments. However, he had an obsolete airways map, and without knowing it, he had flown on the wrong side of the radio range. He circled at 6,000 feet with turbulence bouncing the airplane. The wings were icing.

A surviving passenger later said the airplane dropped several thousand feet in a downdraft, and flew just above the dense forest that dominates the mountainsides of Southeast Alaska. It was dark, and the fog extended almost to the ground. The Electra narrowly missed one mountain. Shortly afterward, another mountain loomed ahead. Gillam, with only one engine, tried to climb the crippled plane over it, but he didn't make it. The right wing hit a tree and the plane plunged to the ground.

The Search for Gillam

The aerial search for the lost plane started at daylight next morning. The Coast Guard and fishermen with their own boats immediately started searching the shorelines around the many islands and inlets near Ketchikan. Both civilian and military planes winged above dozens of channels, bays, and inlets. They had little to go on, other than that the plane was down somewhere more than four hours of flight time for the Electra from Seattle.

Southeastern Alaska is made up of many islands, big and small. Much of it is heavily timbered, with steep mountainsides, and, in January, jagged snow-covered peaks. There is no rougher mountainous country in the world.

After days of searching during the short daylight hours of January without seeing any wreckage, many suspected the airplane had crashed in the ocean, or in one of the hundreds of deep water bays or channels that typify Southeast Alaska. Snow, fog, clouds, rain and sleet hampered the search.

The plane crashed in tall timber, ten miles from a shoreline. It was almost impossible to see from the air. The survivors often heard, and sighted, searching airplanes. Despite waving and signaling, they had never been seen.

The search was called off after only two weeks.

Two weeks later, on February 3, 1943, Coast Guardsmen aboard a patrol boat spotted a bonfire on the beach of Boca de Quadra Inlet, about forty miles from Ketchikan. Investigating, they found two bearded men running up and down the beach, shouting and sobbing.

"Who are you?" one of the Coast Guard crew called.

"Survivors from the M-K plane that crashed more than a month ago," was the response.

They were Percy Cutting, an M-K mechanic, and Joseph Tippets, a Civil Air Administration engineer. Tippets had sprained ankles, and Cutting's back was fractured. They were the least injured of the four survivors.

Twenty-five-year-old Susan Baxter, a C.A.A. stenographer, had died of her injuries within forty-eight hours of the crash.

Cutting and Tippets had waited at the wrecked plane for help until it became clear the search had ended. For nine days they had fought their way ten miles to the beach, down steep icy slopes and through dense brush. Their shoes became so shredded they had tossed them away and wrapped their feet in rags. With an abandoned dory they had tried to row out of the bay, but it capsized and they had to swim ashore

They managed to shoot two crows with a .22 rifle the plane had carried, and they ate half a cup of weavily rice they found in an abandoned cabin. This was the only food they had eaten during the previous nine days. They had each lost at least fifty pounds.

Robert Gebo, M-K's Alaska's general superintendent, and Dewey Metzdorf, an Anchorage hotel man, were the other two surviving passengers. Neither was ambulatory. One had a broken arm and leg, the other a fractured collarbone. They were near death in a crude leanto when rescuers, led by Cutting and Tippets, arrived.

Harold Gillam, who suffered a gashed head in the crash, worked for the survivors organizing shelters, cooking food, tending fires. On the sixth day after the crash he had taken some food and a parachute for warmth, and told the others he was going to search for a landmark of some kind. The others couldn't dissuade him from leaving. They worried about the head gash. But he had left anyway. They hadn't seen him since.

After days of searching, on February 6, a Coast Guard detail found Gillam's frozen body, wrapped in the parachute, only a mile from where Tippet and Cutting had been found. He had hung his red underwear in a nearby spruce tree, and hung his flying boots, upside down on two nearby poles. He probably died from exposure and freezing.

Cutting, the mechanic, found where Gillam had broken through the ice in a stream. He theorized that Harold had hung his clothing to dry, and wrapped himself in the parachute and lay down to rest. When he was found, his injured head was lower than his body. Cutting thought that blood rushing to his head may have caused a cerebral hemorrhage.

There was no autopsy.

In August, 1943, in their report on the investigation of the accident, the Civil Aeronautics Board summarized, "While the stoppage of the left engine from an undetermined cause in extremely rough weather and over hazardous terrain undoubtedly was the primary cause of the accident, it is apparent that strong contributing factors were the pilot's failure (1) to equip himself with an up-to-date aeronautical chart and (2) to utilize the radio aids available to him to accurately establish the position of the flight while on instruments."

Gillam's spectacular flying career had lasted fifteen years.

Funeral services for Harold Gillam were held at the Fairbanks Elks Hall March 5, 1943 with the Christian Science Church officiating.

There was a large attendance and many floral offerings. Pallbearers were James Barrack, of Fairbanks' Samson Hardware; Clyde Smith, a friend; James Dodson, pilot; Noel Wien, pilot; Frank Barr, pilot; and Thomas Appleton, aircraft mechanic. Burial was in the Elks plot of the Fairbanks cemetery.

He is memorialized in Anchorage by Gillam Circle, which lies a few blocks north of Lake Spenard, and near Borland Drive, Will Rogers Place, Wiley Post Loop, and Crosson Circle and Crosson Drive.

At Fairbanks, there is Gillam Way, which passes close to where Gillam's hangar once stood at Weeks Field.

Geographical locations in Alaska named in his honor include Gillam Falls in the Tebay Lakes region, Gillam Glacier, sixty miles East of Healy in the Alaska Range, and Gillam Lake, in the Mentasta Mountains of the Alaska Range.

He was survived by four children; Harold Jr., Donald, Maurine, and Wenona.

17 Ed Young

BY WARREN O. TILMAN, AS TOLD TO JIM REARDEN.
(See end of this chapter for background on Warren O. Tilman)

I met Ed Young at Anchorage in the winter of 1930-31, when he flew for Alaskan Airways. I had heard of him, of course; he had flown to Siberia on the Ben Eielson/Earl Borland search, and he had flown the Fairchild 71 carrying their bodies back to Fairbanks. Like others involved in that event, he had received much publicity. It was at that time I also met Noel Wien, Mat Nieminen, Joe Crosson and other pioneering Alaskan pilots.

I moved from Anchorage to Fairbanks in April, 1931, and went to work for Alaskan Airways, at which time Ed Young and I became good friends. Soon I became his flight mechanic, and I flew on many trips with him, as was the custom then. Mechanics were called "mechanicians" and commonly accompanied pilots on commercial flights. Especially in winter there was a much work involved in caring for an airplane overnight, plus, airplanes needed much mechanical attention in those days. He liked the way I maintained and repaired aircraft; and I liked to fly with him, for I thought he was a very safe pilot. I repaired a lot of airplanes in those years, but I never had to fix one that Ed Young broke.

Young wasn't a big man, but he was husky. He gave the impression that he could take care of himself whatever came along. He made a lot of friends, for he was thoughtful of others. He was very independent, with a mind of his own. He was always in complete control while flying.

During World War I, Ed Young learned to fly in Jennies at San Diego, California, while he was in the Army. He was commissioned a second lieutenant and became a flight instructor at Payne Field, Mississippi, after which he was sent overseas where he was taught how to fight a pursuit plane. He had completed the course and was about to be sent into combat when the war ended.

Pilot George Ed Young as portrayed by Alaskan artist Harvey Goodale.
Courtesy of the Bob Reeve Family

He then became engineering officer for the Twenty-Second Aero Squadron in the Army of Occupation. Following that he was stationed at Coblenz, Germany with the 638th Squadron.

At war's end, he, with other pilots, flew Spad fighter planes from the factory to a field in France where the engines were pulled, and the airplanes were burned. Since the engines wouldn't burn, he presumed they hauled them to a dump. Daily, he picked up a new Spad, flew it to the field, and returned by train to do it all over again.

"Spads were fun to fly. It was one of the better planes of World War I. At least their wings didn't fall off in a dive, or the fabric didn't peel, as was true of a few of the others earlier. They were fun to dive full throttle; on the pull up, the momentum they built up gave them a great zoom," he told me.

He left the Army in 1919 and returned to his home in Michigan, where he flew as a commercial pilot for a time. He arrived in Alaska in 1924, with 1500 hours of flight time in his log book, and lived in the Lake Minchumina area near great Mount Denali where he hunted and trapped. No one in Alaska knew he was a pilot until May, 1925, when Noel Wien was forced down in the Kantishna country, forty miles southwest of Nenana, and Jimmy Rodebaugh needed a pilot to fly a search for him.

Ed was in Fairbanks at the time and volunteered to fly the search. He test hopped the second Standard the company owned (Wien had disappeared with the other). The engine overheated and ran rough, and he landed. Ralph Wien hauled out Eielson's old Jenny and worked on it to make it airworthy, and Ed flew it around a bit and landed. He was not familiar with a soft spot on the runway (Eielson's soft spot, where his mail plane flipped), and on landing he found it and the Jenny nosed over, cracking the prop.

Rodebaugh hired him as a mechanic, with the possibility of his becoming a pilot, and Ed never went back to his Minchumina trapline. Once he started flying, he pioneered Alaska's skies with the likes of Noel Wien, Joe Crosson, Ben Eielson, Harold Gillam, Anchorage's Russell Merrill, and others.

Alaskan Airways won a winter contract to fly the mail once a week from Circle City to Fort Yukon, and Ed did the flying with me as his mechanic. At the same time, we flew to Eagle every two weeks, and to Beaver from Fort Yukon once a month.

We usually flew a Fairchild 71, a high-wing monoplane, a good plane for those days. It performed well with a 420 hp Pratt & Whitney radial engine. Wingspan was fifty feet. The wings folded back against the fuselage when it was stored. A vertical bolt on the front wing spar hinged the wing to the fuselage. A horizontal tapered pin with a handle and a spring held the rear spar to the fuselage. A short flap on the trailing edge of the wing was raised, and a light jury strut steadied the wing when folded. This was a simple and safe arrangement and very handy. I remember putting three Fairchild 71s in a sixty-feet-wide hangar that Alaskan Airways had bought from Noel and Ralph Wien.

The Fairchild carried seven passengers, including the pilot, plus 160 pounds of baggage. Its three fuel tanks held 155 gallons of gasoline. Cruise speed was about 100 mph, with the engine using twenty-two to twenty-five gallons per hour.

A ticket to Nome from Fairbanks was $500; to Ruby or Wiseman $250. Parcels cost thirty-five-cents a pound. It wasn't long before the passenger fares were cut in half, as more and more people flew. However, the parcel price remained the same.

Skis used on these planes for winter flying were eleven and a half feet long, eighteen inches wide, and weighed 165 pounds *each.* Fairbanks carpenter Charlie Schiek made them of steamed and bent hardwood. The Fairchild company made the ski pedestals, but they weren't much good until Jim "Hutch" Hutchison strengthened them by welding on more tubes. The tail skid was a big steel shoe, on the bottom of which Hutch welded some stellite (extremely hard metal) for better steering.

During winters, each week Billy Root hauled the mail by automobile from Fairbanks twenty miles on the Steese Highway to the small town of Chatanika. From there, Johnny Palm and his crew hauled it with horse-drawn sleighs the 142 miles to Circle City. What a story that tells of the 1930s Alaska – horse-drawn freighting sleighs to airplanes in one jump.

After a big storm that dumped a lot of snow, on one flight to Circle City we flew over Palm and his men with their sleighs and horses. They were dug in, and hadn't moved for several days. On our next trip, they were still there. Of course we had no mail to haul from Circle City that day.

When we flew over them a for the second time, with them in the same place, Ed remarked, "Maybe we're in the right business after all."

Before we left for our weekly trip to Circle, all the mechanics – Jim Hutchison, Gordon Springbett, Jack Warren (who ran the Teller roadhouse from which Eielson and Borland departed on their fateful flight), and Orville Porter (the mechanic who accompanied Wilkins, Eielson, and Crosson to Antarctica) plus Hjalmar Nordale, one of the company officials, lined up to shake our hands before we boarded the Fairchild.

"They act sort of like they don't expect to ever see us again," Ed once commented.

The same ceremony took place when we returned – they all ganged around and each would shake our hands when we got out of the plane.

Hand shaking also seemed to be part of landing in villages in those years; whenever we arrived at Circle City, Fort Yukon, Eagle, or other village, the people surrounded us and all shook our hands. It seems a bit strange today, but at the time it seemed to be a nice ceremony.

Weeks Field had a short runway, and in winter on takeoff we were concerned that we might plow through the Northern Commercial's cordwood pile at the east end. Ed had to chop the power many times when he saw we weren't going to make it. As that big radial engine started to roar and we started down the runway I always wondered, "Will we, or won't we…"

When the flight was aborted, to lighten the plane sometimes we'd drain a little fuel, or leave a little freight. Seldom did Ed offload the flight mechanic.

When we arrived at Circle City, Ed buzzed the village to alert an old man with a horse and sleigh to bring us the mail for Fort Yukon, and we'd land on a lake a few miles to the west. It often took a long time for the horse to arrive, and we finally engaged Silas Gunnison with his dog team to do the job, a much better arrangement. He could harness his dogs and speed down the trail. The ride in his sled back to the village when we stayed overnight could be a little rough, but we were young enough that it didn't bother us. Once Silas's sled overturned, and the dogs and the mail and I rolled down the bank of the Yukon. I was unhurt.

If the Yukon River froze up smooth at Circle City, we'd land close to the village, which was more convenient than landing at the lake.

Circle City had perhaps seventy-five residents, a Northern Commercial Company store run by Charlie Alexander, and a sawmill that made lumber from winter-cut logs. There were also a few nearby gold mines. The Romakers had a two-story roadhouse where ten or twelve guests could be accommodated. Jack Barton was the Army wireless operator for the Alaska Communication System; old Tom McClain was the marshal.

The Romaker roadhouse was gradually sinking into the permafrost, and the floor and the log walls were all cattywampus. One evening, Mrs. Romaker handed us a kerosene lamp, and while Ed and I were climbing the crooked stair steps to the second floor bedroom with it, he commented, "Tillie, I don't mind sleeping in these old log roadhouses in the winter. I think the frost might hold the logs together. But in the spring thaw it might be another story."

On one trip to Circle City, Ed landed on the little lake and taxied to the shore near the trail. It was very cold, and we walked around a bit to keep warm while we waited for the old man and the horse sleigh to bring the mail sacks for Fort Yukon. A tall, heavy set bruiser, driving what looked like a pack of harnessed wolves, drove across the ice and stopped near the Fairchild. He had probably just come off his trapline. He was about the meanest looking man I have ever seen. He wore a long black fur coat instead of a parka. He had a big black, villainous mustache, and he frequently cussed his dogs in a rough, loud voice. They appeared to be afraid of him, and I didn't blame them. An Indian girl, about 15, wrapped in furs, lay in his basket sled. She had a pretty face, but she looked very ill.

"The girl is my daughter, and you're going to fly her to Dr. Burke at the Fort Yukon hospital," he roared.

He acted as if he owned the airplane.

Ed faced him down. "The girl will fly to Fort Yukon only if I say so," he said, calmly.

That set the trapper off. He yelled at Ed. "You will take her to Fort Yukon, right now," He had a long whip, which he popped a couple of times like rifle shots to emphasize his ultimatum. I eyed the pick handle on the sled with which the big guy controlled his dogs, wondering if he might try it on Ed. I was about ready to hunt a log to crawl under.

Ed didn't bat an eye. He made it clear that he flew the airplane, and made all the decisions regarding it.

I didn't think the guy would back down to Ed, but he did.

The atmosphere changed when we all realized Ed was in control. There was still the matter of a fare, and Ed refused to take the fur the guy offered. He finally dug out a big roll of bills and paid cash.

To make room for the girl so she could lie down, I took two wicker seats out of the back of the Fairchild's cabin and put them in the lakeshore brush. I'll bet they're still there.

Within another hour or so we unloaded the girl into another dog sled at Fort Yukon, which sped her to the little hospital. I never heard how she made out.

I remember another incident when we landed at Circle Hot Springs on our way home from our upper Yukon River mail trip. It was a sunny, pleasant, mild day. Old Pop Leach showed up on the runway driving a team of horses pulling a sleigh. With him were two young women and two young men who wanted to fly to Fairbanks with us.

One of the women had a camera and wanted to take a picture of the others as they stood by the ski on the left side of the airplane, directly behind the wing strut. Ed had started the engine, which was quietly idling. He had me stand close behind the door to watch that no one would get excited and be hurt getting into the plane. The girl with the camera stood by Ed in front of the airplane to take a picture of the others.

She stepped closer to the subjects, and didn't notice the prop when she quickly moved past Ed. Suddenly she was within inches of the spinning prop. Ed was a quick as a cat. He grabbed the arm of her parka and jerked her to safety. In about one more second she would have been girlburger.

We usually didn't start the engine until all the passengers were seated and their safety belts fastened. That was a close call, and Ed commented later that we had been foolish to allow it to happen.

I don't think I've ever seen anyone move as fast as Ed did on that occasion; his arm was just a blur.

The seventy or so miles from Circle City to Fort Yukon usually took us less than an hour. At Fort Yukon we stayed overnight with Sergeant Curlee, the wireless operator for the Army Signal Corp. When we landed, I got out and shoved spruce poles in front of the skis, and Ed taxied up on them. This kept the skis from freezing down overnight.

If we were going to spend the night, as soon as the propeller stopped turning I drained the twelve gallons of engine oil and kept it warm overnight on a wood stove in Curlee's cabin. I then refueled the Fairchild with case gas we had stored at Fort Yukon.

When we stopped the engine we had to run the carburetor dry. If we didn't, when I pre-heated the engine for the next start, gas in a full carburetor could expand, overflow, and run into the heating firepot, which, of course, could flash into flames that could destroy the airplane.

The big canvas cover that went over the engine hung clear to the ground, and there was room to crawl into it while heating the engine. Even in the coldest temperatures it was impossible to remain very long under the tarp because of the fumes from the gasoline-burning firepot. It was nice and warm for as long as I could stand it.

Fort Yukon was an Athabascan Indian village, with only a few white residents – several traders and their helpers, Curlee, the Signal Corp guy, a game warden for the Alaska Game Commission, a preacher or two and a few retired white trappers.

Trapping of marten, fox, beaver, muskrat, mink, otter and wolf, and catching salmon with fish wheels from the Yukon River for dog and human food, were the two major industries. A few Indians cut cordwood they sold to sternwheel riverboats in summer.

Because we usually remained overnight, Fort Yukoners were

able to answer their letters we had brought during the night we were there, and we'd take them with us when we left next morning.

Nights after supper at Fort Yukon, Circle and Eagle, to prepare the field for morning takeoff, we often went out and snowshoed back and forth in front of the airplane. Usually, villagers went along to help, for they appreciated air service. Ed and I didn't mind the snowshoeing. Often it was clear, with frost in the air, and sometimes there were spectacular northern lights. Occasionally, we'd hear wolves howl.

In the morning I'd preheat the engine with the plumber's firepot we carried, clean frost off the wings, and pour the hot oil back in and start the engine. Then we'd fly on to the next stop with a couple of sacks of mail, and maybe a passenger.

It sounds crude now, but in those days people thought it was wonderful. And, by damn, it was, compared to dog team passenger and mail service. Small wonder it took but a few years for the transition from dog teams to airplanes to haul bush mail and passengers in wintertime Alaska.

When we made the run to Eagle, as we flew up the Yukon River we sometimes landed at Woodchopper or Coal Creek with a little mail or a package. Sometimes, if we saw someone we recognized, instead of landing we'd just fly by and throw a mail pouch out to him.

We occasionally landed at the mouth of the Nation River and put mail in a box on the bank of the Yukon for "Phonograph" Nelson who lived nearby. Once we landed there on glare ice and the wind blew us for a mile along the north bank, turning the plane around and around. Thanks to Ed's ability with the Fairchild, we missed the rock bluffs, and the mid-river rough ice.

Before taking off at Circle City, Fort Yukon, Eagle or wherever, Ed usually walked along the runway looking for obstacles and checking the wind.

Once at Eagle, there was such a strong crosswind where we usually landed and took off that Ed couldn't consider taking off as usual from the old parade ground that had been used by soldiers stationed there at the turn of the century.

"What do you think, Tillie?" he asked, as we walking along, occasionally staggering from the wind.

"You're the pilot. But you might consider taking off from the wireless station and across the old parade ground," I suggested.

Everyone in Eagle, and their dog teams, helped get the airplane up the hill there. Many dogs pulled, and people pushed and pulled, and all had a grand time.

We pointed the nose of the plane toward a row of cabins on the far side of the parade grounds, cranked the inertia starter (a flywheel, that when hand-cranked to full speed, was then engaged to turn the engine over), got her started and nicely warmed.

When Ed opened the throttle I learned that an airplane could sure accelerate when going downhill. Facing the wind, and with only a few sacks of mail, we had no difficulty clearing the cabins, although the roofs looked close to the skis when I looked down.

Once when we took off from Circle City flying the thirty miles or so to Circle Hot Springs, there was so much headwind that we climbed to 11,000 feet, which was easy to do with a Fairchild 71 with no load. We could still see the Yukon River, as well as Circle Hot Springs.

We had been flying for more than thirty-five minutes, and we were only half way. "Seem's like we're just painted into the sky here," Ed said. He descended, and from the sawmill at Birch Creek to Circle Hot Springs we flew just above the tree tops where the air was smooth, despite the wind. I counted sixteen moose in that short distance.

I always enjoyed remaining overnight at Eagle. We stayed at Mr. and Mrs. Bob Steel's roadhouse where we had good beds and good food. The Steels later moved to Fairbanks with all their kids, where they ran the Steel Hotel on Front Street, near Cushman Street.

Tiny Knight was the signal corps wireless operator at Eagle. He had a huge dog team; he needed all of them, for he was a big fellow and needed all that power. Charles Thompson and Horace Beederman both had trading posts at Eagle.

Nimrod, a colorful Eagle resident, was the local dentist. When he drilled a tooth on a patient, the patient had to turn a handle for a flexible shaft that allowed Nimrod to do the drilling. Once Nimrod took me to his place to show me a new way he had for drilling teeth; he had installed a drill chuck on the works of a big old alarm clock, into which he could clamp his drills. He could wind that old spring up, and drill away.

Old John Powers, a Territorial Senator and his wife, Ma Powers, lived at Eagle. They made a great home brew. She cussed fully as fluently as did old lady Quigley over in the Kantishna.

At Eagle we dropped off the mail for the Fortymile, and an old fellow there took it over the trail with two horses, that is, if he could find them. He always wanted us to fly him around to try to locate those two old nags.

We were once stuck in Eagle for a week by bum weather, and I enjoyed every minute. I spent time with Barney Hanson, a prospector who for years was my close friend. We went a little way out of town and shot at targets, and we hitched up his dog team and went upstream the four miles to visit the Indian Village.

Ed spent much of that week at the roadhouse trying to put together a puzzle ring he had bought from Nimrod. He also bought a knife from Nimrod for ten bucks – Swedish steel, a horn handle, and Nimrod had inlaid a quarter ounce gold nugget in the handle. Gold was $20 an ounce then.

Charlie Thompson had two Native-made caribou skin parkas for sale at his trading post. Ed bought one for $17, and I'd have bought the other, but I didn't care for the fancy colored beadwork on it. Instead I bought some beaded gloves.

The local Natives at Eagle sold a lot of their handwork to Thompson, and much of it was decorated with glass beads. It was beautiful work. He offered me a moosehide belt that was about ten inches wide, with a jillion pretty beads in beautiful designs. It was made to go across a mother's head and under a baby's bottom for carrying. The belt was endless. Stretched, it must have been about six feet long. The leather side was dry tanned and as soft as flannel. Charlie wanted ten bucks for it. Today it would be priceless. I turned it down.

Ed was always prepared for emergencies and often commented, "If the engine quit now, here is what I'd do. See that good smooth bar there, it would be a good place to land. There's a cabin up there a few miles where we could survive nicely. This is good game country and we could live off the land if we lost an engine."

Once in the springtime, with ice still on the rivers, we were at five or six thousand feet and it was clear and sunny. "Do you think we could land on that creek? " he asked, pointing down. To me it looked straight, but narrow and short to land on.

"Watch," he said. He then flew so the shadow of the airplane followed along the center of the straight section of the creek. The snow-covered creek ice was at least three times wider than our shadow. At an indicated ninety miles an hour it took over sixty seconds by my watch to cover the distance of the straight section. We had kept our altitude, but Ed had proved to me that we could have easily landed there if need be.

He was a precision pilot. I watched how he handled his landings. When he arrived where he was to land, he maintained altitude and flew over the spot and gave it a good look. He then circled and flew a left pattern and descended without slipping the airplane or gunning the engine, and he always greased her down. He never fell short and had to blast the engine, and he never overshot.

After we had landed on the hard snow at Circle Hot Springs all one winter, I could just about count every one of our landings by the tail-skid tracks at the end of the runway. They were so close together I could have stood in one spot and almost spit on each track where he had touched down.

Once while we were flying up the Seventy-mile River from Eagle the wind was blowing south over the mountains from the Yukon so hard we could hardly climb. We had to fly north, and get over mountains to reach Circle City. We were flying Fairchild NC153H, which had a freshly overhauled engine. (Fairchild 71 NC153H was the funeral plane in which Young flew the bodies of Eielson and Borland from North Cape to Fairbanks).

Ed gave the engine full power and set the stabilizer for the best climb. He told me, "Tillie, if we ever get over the top of this ridge, we'll really climb on the other side." Just as we crossed over the ridge, oh my how we went up. Ed just throttled back and nosed down and took it easy. It was just as he had predicted; we climbed like a balloon. We just about coasted all the way to Circle City.

Ed excelled at flying in bad weather. Once, after leaving Circle City and headed for Fort Yukon, the wings started to pick up ice. Soon he had to use full power just to keep us in the air, and the old Fairchild was wallowing. We were within five minutes of Fort Yukon when we had to turn around and head back to Circle. We were soon out of the bad stuff, and we landed safely at Circle, where we remained overnight, and made the flight to Fort Yukon the next day.

Ed knew a lot of the trappers and prospectors who lived in the bush. I don't know how he kept the names and places straight, but he did. We often flew out of our way to drop things to these remote-living people. We always circled their cabin to get them to come out so they could find what we dropped for them.

Ed had a great knowledge of the country, and what was safe and what wasn't. I think he learned a lot of this when he trapped for several years in the Minchumina area. When we wanted to land on a strange lake for the first time in winter, he would touch the skis down on the snow and give the plane full power and fly across the lake with the tail in the air. He pulled up after crossing the lake, and circled to check the tracks he had left. If there was overflow, the tracks would be dark with water – not a safe place to land. At times a heavy load of snow will push lake ice down, and overflow water will cover the ice. Deep snow over overflow will keep water from freezing for a long time.

Ed Young was safe for many reasons. He wouldn't fly an airplane that needed attention. One day he left Fairbanks flying freight to Livengood with a five-passenger Stinson biplane. He returned in a few minutes and told us the airplane had no compass.

He could have flown to Livengood without a compass, no problem, probably with his eyes closed. But he always played it safe. We shifted his load to another airplane with a compass so he could make the trip.

We carried good emergency equipment in the Fairchild. There was always a chance of a forced landing, whether it be weather, a mechanical problem, low fuel - whatever. Once down we would have to depend on that emergency stuff for survival.

In the baggage compartment we carried a gun and ammunition, axe, rope, and two pairs of large snowshoes, along with the engine cover and a fire pot to heat the engine and two seven-and-a-half-gallon oil cans for draining the oil each evening as soon as the prop stopped.

We also carried spare wool shirts, underwear, pants, wool socks and spare felt insoles, spare wool mitten liners, canvas gloves, plus water boots and spare sheepskin or caribou fur socks.

We had pots and pans, salt, matches in waterproof containers, snare wire, and some fishing tackle, a large can of tallow, flour, tea, coffee and chocolate, sugar, dried eggs, raisins or prunes.

For a time we carried an over and under Marbles game-getter, with a folding metal stock. It had a .22 caliber barrel and a .44 caliber barrel. We always had a box of .22 long rifle cartridges, a box of .44 caliber ball cartridges, and a box of .410 shotgun shells. Ed didn't have much faith in this little gun; he was much happier when I brought along a .30-30 Winchester carbine, with ammuni-

Ed Young in typical Alaska bush pilots attire, posing on the skis of a Fairchild 71. Young was a military pilot in France during World War I. He was a pioneering pilot in Alaska during the 1920s and into the 1930s. Circa 1930.

tion. He felt we would have a much better chance to get a moose or caribou with it, and I think he was right.

In 1932, Pacific Alaska Airways, (a subsidiary of Pan American Airways) bought out Alaskan Airways, and in 1933 they started to build a new hangar. I didn't want to build a hangar, so I took the summer off. I did some work at Weeks Field in the hangar of Percy Hubbard and Art Hines, and painted a few cars just to make a few bucks and keep busy.

Ed Young was named Maintenance Engineer for P.A.A. He asked me to go back to work for P.A.A. in the newly finished hangar. I agreed to this, and he hired me then and there. Next morning, at 8 a.m., I arrived at the hangar as an Aircraft and Engine mechanic.

We loaded two Fairchild 71s with freight for Livengood that morning, and Ed Young and S.E. "See" Robbins soon flew them off. Ed also had two passengers.

About an hour later Charley Burgenson came to the hangar and told me that Ed and his two passengers had crashed at Livengood and all three had died. It was one of the toughest blows of my life; I felt as if I had lost my best friend.

Robbie had landed and pulled off the Livengood runway to make room for Ed to land. He saw what happened, and described it for me later. He said Ed had just turned from his downwind leg and was starting to turn for his final approach to the runway, but his plane kept turning and went straight down.

Sam O. White, a flying game warden at the time, and later a much-admired bush pilot, said, "Tillie, if Ed Young was killed in an airplane, the rest of us had better quit flying."

Sam, like me, thought Ed was the safest pilot flying in Alaska. Frank Dorbandt, another local pilot, once told me that he thought Ed Young was the best all-around pilot in the Territory.

I was sure it wasn't Ed Young's flying ability that caused the crash. At the time all the pilots got a monthly physical examination from old Doc LaVern. Most of the pilots I knew lived good clean lives, and I doubt that the old Doc gave them a very complete exam.

Frank Pollack, a prominent Fairbanks pilot of the time, had recently flown to the Quigleys in the Kantishna district with mining supplies. Fannie Quigley told Frank that Ed Young had landed there a few days earlier, and had visited for a time.

"He looked sort of funny, as if he was going to pass out," Fannie told him.

She tried to get Ed to stay, but he sat and rested for a short time, and said he was all right. He seemed to feel better when he left.

Fannie told Frank she thought Ed had a heart problem, and she believed he almost passed out during his visit.

Hjalmer Nordale and a couple of other P.A.A. company people went to Livengood to inspect the wrecked Fairchild. "All of the controls were intact, cables, pulleys and all. We could find nothing wrong with the plane," he told me.

Noel Wien flew the bodies back to Fairbanks. When they arrived at the hangar, Sourdough Express brought their truck to haul them to undertaker Hosea Ross. Coal had been hauled in the truck, and there were still a few pieces in the bed. I wouldn't let them load the bodies until I had swept the truck clean.

Sam O. White, like many of us, loved Ed Young. "Fairbanks needs some changes," he said. "I hate to see bodies of our people hauled in a coal truck."

The next day he started a campaign to raise money for an ambulance. Enough was quickly raised, and the new ambulance was kept at the fire hall.

We had a funeral, and buried Ed below the sawmill in the old cemetery near the Chena River. Just about everyone in town attended. In later years, pilots Art Hines and Harold Gillam were buried close by.

WARREN O. "TILLIE" TILMAN

Aircraft mechanic and registered Alaska big game guide Warren O. "Tillie" Tilman with whom I wrote the previous chapter, was born in the Missouri Ozarks in 1906, and became involved with airplanes as a young man. Over the years he often brushed shoulders with history makers while he worked on their airplanes. His interest in airplanes started when he was 5 when he watched the famous Gin Fizz, *a tiny airplane attempting to span the United States, when it flew over his then-home-town of Parsons, Kansas.*

In 1928, he worked in the Travel Air airplane factory at Wichita, Kansas, installing fabric on biplanes, and getting acquainted with owners of the factory Walter Beech, Clyde Cessna, and Lloyd Stearman.

In 1929 he worked for Boeing at Seattle, and helped to crate two of the three Fairchild 71 airplanes being rushed to Alaska to join the search for Ben Eielson and Earl Borland.

He arrived in Anchorage in 1930, and moved to Fairbanks in 1931 where he worked as a mechanic for Alaska Airways. With mechanics Jim Hutchison, and Eddie Moore, Tilman serviced the Winnie Mae, the Lockheed Vega of Wiley Post during Post's and Harold Gatty's 1931 round-the-world flight.

Tillie worked for Pacific Alaska Airways at Fairbanks after that organization bought Alaska Airways. During those years he knew and worked on the airplanes of such notables as Joe Crosson, Noel and Sig Wien, Harold Gillam, Ed Young, Frank Dorbandt and many other early Alaska pilots.

Warren O. Tilman, longtime aircraft mechanic, and registered Alaskan hunting guide. He flew with Ed Young as a mechanician and knew and worked on the airplanes of Wiley Post, Amelia Earhart, Noel Wien, Joe Crosson, Harold Gillam, and many others.
JIMMY BEDFORD

In 1935, at Juneau, he repaired the starter of Wiley Post's airplane while Post and Will Rogers visited locally. By the time those two travelers arrived in Fairbanks, Tillie was also there. Will Rogers planned to make a fishing trip with Tillie after he and Post returned from Siberia where the two planned to visit. That fishing trip, of course, never took place; Rogers and Post died when their airplane crashed near Point Barrow.

Tillie was again employed at the Seattle Boeing plant and, later, at the Lockheed plant at Los Angeles, where he worked on the Lockheed Electra of Amelia Earhart.

During World War II he worked for Boeing at Seattle, helping to build B-29 Boeing Superfortress bombers. After the war he returned to Alaska, and when I (author) first knew him, he worked for Northern Consolidated Airlines at Fairbanks. Among those he guided for big game in Alaska was General Jimmy Doolittle.

He left Fairbanks in 1962 and established a one-man shop in Denton, Texas, where he custom-built aerobatic airplanes, completing fourteen Pitt Specials, and fourteen Stardusters. He retired at age 79 and returned to Fairbanks.

I met Tilman in 1950, when I lived at Fairbanks, and we became friends. I hunted and fished with him, and I wrote and published a number of his hunting yarns. We carried on a correspondence for about forty years. In April, 1981, he sent me a carbon copy of an eighteen-page, single-spaced, typewritten letter he had written to his four sons. Much of the letter was about his friend, Pilot Ed Young. Today, Young is largely forgotten even in Alaska.

I condensed the Ed Young part of the letter, and with Tillie's approval, the material was printed in the Fairbanks Daily News Miner's *Sunday* Heartland Magazine *of May 21, 1989.*

Warren O. Tilman spent his last years in the Pioneer's Home at Fairbanks. He died in 1994 at the age of 88.

Chapter 17, **Ed Young**, *is a more detailed version of Tillie's letter than the article that appeared in the* News-Miner.

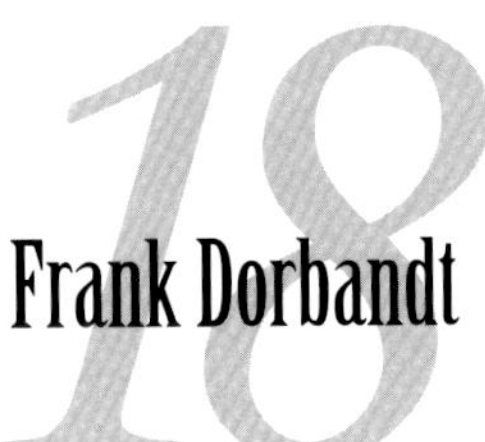

Frank Dorbandt

EARLY ALASKA BUSH PILOT Frank Barr remembered meeting Frank Dorbandt in 1926 in Detroit, Michigan. Barr, with other young pilots flying from Triangle Field was impressed by the six-foot-tall, blond, burly extrovert who was there to pick up an airplane. He entertained the young flyers with hair-raising tales of his flying exploits.

Barr met Dorbandt again in 1933 at Seattle. Barr was now a pilot for Bill Strong, a trader at Tulsequah, on the Taku River about forty miles from Juneau. He was in Seattle to pick up a Fokker amphibian which he was to fly commercially for Strong from Juneau into British Columbia and the Yukon Territory. He innocently told Dorbandt about plans for the Fokker.

When Barr landed in Juneau with the Fokker, his boss handed him a telegram that had arrived from Dorbandt. In it Dorbandt detailed the reasons it would be foolhardy for Strong to use a pilot with limited Northern experience (Frank Barr) for the Fokker operation. Dorbandt offered to replace Barr. This generous offer from Dorbandt didn't endear him to Frank Barr. (Dorbandt didn't get the job).

Barr's next encounter with Dorbandt came when the engine of the Fokker amphibian died and was sent to Seattle for repairs. While waiting for the engine repair, Barr flew from Atlin an English-built two-cockpit Moth biplane. It was a small, seventy-six-horsepower airplane, hardly adequate for commercial use, but it was all that was available.

Barr, at Atlin's Kootenay Hotel, was awakened one May

Don Glass (left) and Frank Dorbandt (right) flew this Ford Trimotor from Seattle to Alaska in 1934. There were eight passengers and two pilots, plus much freight in the big plane. The Ford was frequently low on gas, and landings were made in places where no plane had ever before landed.

morning in 1934 by Don Glass, a medium-sized, dark-haired, friendly pilot.

"Can you fly some gas to our plane? We ran out of fuel with a Ford Trimotor and landed on the ridge between Gladys and Surprise Lakes. I'm the copilot."

"Where are you headed?" Barr asked.

"Left Seattle a couple days ago. Headed for Anchorage," Glass answered.

"Okay. But I can only haul thirty gallons of case gas in the front cockpit," Barr answered. "Who is the pilot?"

"Frank Dorbandt," was the answer.

"Not interested," Barr said, and rolled over to go back to sleep.

Glass explained there were passengers who weren't properly dressed, and that there wasn't much food. It was a thirty mile hike out of the mountains for them to reach Atlin.

Since passengers were involved, Barr couldn't refuse to help. He flew six five-gallon cans of gas and some sandwiches to the Ford, where he found seven miserably cold passengers and Dorbandt sheltering from an icy wind next to the downed plane. Dorbandt had landed on a short grass, treeless, mountain meadow that was reasonably smooth. He excelled at handling an airplane under difficult conditions.

The "angel," a 24-year-old from Los Angeles named Marshal Kester, who had backed Dorbandt with dollars to buy the Ford, was one of the passengers. He weighed at least 300 pounds and could never have walked the thirty miles to Atlin. The others, mostly young, poorly shod for mountain walking and lightly dressed, weren't much better off.

All, including Dorbandt, welcomed Barr enthusiastically.

Dorbandt unloaded the freight and baggage, loaded the passengers, and flew to the Atlin airstrip. He soon returned for the freight and baggage.

Barr wondered how an experienced pilot had managed to run so short of gas he had to make an emergency landing in the mountains. Don Glass explained, that after a roistering night at Telegraph Creek, Dorbrandt needed a nap after a morning take off. "Take the controls. Hold this course until you hit seventy-five-mile-long Atlin Lake. Then wake me," he told Glass.

Although Don held the compass course as directed, a crosswind drifted the big Ford sixty miles east to Teslin Lake. Dorbrandt took control, realized where they were, and flew west toward Atlin. When gauges indicated fifteen gallons of gas left, he decided to land before the engines quit.

That night at the hotel, Barr noticed Atlin residents treating Dorbandt coolly. He soon learned why. Two years earlier, Paddy Burke, a pilot flying out of Atlin, didn't return from a late fall flight to the Liard River country. Dorbandt was at Atlin overnight on a flight from Seattle to Anchorage. His employer told him to help on the aerial search for Paddy.

Several planes were involved in the search. More than a month after Paddy left Atlin, all the searching fliers met and agreed there was practically no hope for Paddy because of severe weather and the scanty supplies he had carried. Nevertheless, they all agreed to search one more day.

Instead of searching for Paddy that day with the other pilots, Dorbandt flew straight through to Anchorage.

Paddy was a jolly Irishman, beloved by Atlin residents. When he was finally found dead of starvation and exposure, Dorbandt's name became mud. Perhaps, reasoned the residents of Atlin, his plane on that last day of searching could have found Paddy alive.

Dorbandt had a very high opinion of himself. He was most interested in Dorbandt. He considered himself *un macho hombre*, and proved it whenever he could. One method he used was to fly aerobatics—loops, wingovers, stalls, spins - when arriving at a destination, to the consternation, or fear, of his passengers. Once a passenger claimed the only audience Dorbandt had for a series of aerobatics were a few Eskimos camped on the tundra.

The name Frank Dorbandt became well known in Alaska first when he was involved in the Eielson/Borland search; afterward he became known for his great ability with an airplane. According to Frank Barr, strangely, his name wasn't Dorbandt, but it was the name of an uncle, who had never been to Alaska. His true name was Floyd Andrew DeHayes. Why he changed his name is unclear.

That May, 1934, Seattle-Alaska flight by Dorbandt, during which Frank Barr hauled gas to him, was detailed in a 23-page double-spaced, unsigned typewritten account by a 25-year-old college student who was a passenger on the flight. A copy of the account reached Richard Wien via Martin Hale, a Montana friend, who had received it in the mid-1980s from the former student. Wien passed it on to me (author). It reveals a reckless, carefree, devil-may-care and irresponsible Dorbandt, who was also a highly skilled pilot. The following is a condensation of the account. Since the account is unsigned, I've called its author "Hunter."

Arriving at Seattle, Hunter, who had booked passage to Seward with the Alaska Steamship Company, learned a strike had halted all sailings. A porter at the Olympic Hotel found him a ride to Alaska for $150 in the Ford Trimotor flown by Frank Dorbandt.

Hunter found the trimotor at Boeing Field. Ten people were to make the flight, including pilot Dorbandt and copilot Don Glass, a retired Navy pilot. The aisle between the seats was filled with diesel engine crankshafts, boxes of parts, and other Anchorage-bound freight.

Dorbandt flew the big plane from Seattle's Boeing Field to Vancouver, British Columbia, where it cleared customs and the passengers spent the night at a small hotel. At supper, Dorbandt predicted an arrival at Fairbanks late the next evening, pointing out that in May daylight was almost perpetual in the Far North.

Takeoff next morning was at 9 o'clock. With Dorbandt at the controls, the Tin Goose headed up the Fraser River gorge, bound for 400-mile-distant Prince George.

Around 1:15 p.m. Hunter was startled to realize the airplane was circling a meadow as if to land. There was no sign of an airport, or civilization. To his amazement Dorbandt landed in a wet meadow next to a lake. When the plane stopped it was stuck in mud.

They were virtually out of fuel, and could not have reached Prince George without more gasoline.

They had no camping equipment, no food, and no transportation, and the plane and passengers were stuck on a meadow about ninety miles west of Quesnel, British Columbia, the nearest town.

Clearly, Hunter wrote, flight planning had been almost "criminally lacking."

The improvised landing field was next to a wagon road passable only by horse-drawn wagons or buggies. It led fifteen or twenty miles to an automobile track. Dorbandt and Glass left on foot to seek help and gasoline. Passengers remained with the ship. Hunter was lucky to have a bedroll, hunting clothing, and a rifle. Passenger Frank Williams, an old-time Alaska trader from St. Michael, and his three companions, also had bedrolls.

An elderly French Canadian and his Indian wife, who survived by trapping and hunting, lived nearby in a small cabin. They shared dried moose meat, potatoes from their garden, dried beans, and bacon grease, with the adrift passengers.

The rain, which had been falling when they landed, ended. The eight passengers found a nearby haystack where they spent a satisfactory night.

Two days later, Dorbandt and Glass returned with two barrels of aviation gas hauled in a wagon by local resident Pat Knight and his team of horses.

Everyone pitched in to cut jackpine poles to make a corduroy road in front of the plane. They pried the wheels out of the mud and got poles beneath them. Dorbandt and Glass flew the plane off. The short corduroy road allowed the ship to start rolling before it hit soft ground.

The passengers waited while Frank flew to Prince George to unload the freight and baggage, refuel, and return to pick them up. About 2 p.m. the Ford landed and the passengers climbed aboard. The take off was aborted; the plane couldn't get flying speed. It was again stuck in mud. Once again the wheels were pried up and poles put under them. Half the passengers got aboard, and Frank managed to get into the air and fly to a small hard-ground hayfield ten or fifteen miles away where some Indians were holding an amateur rodeo. He landed there and left to retrieve the rest of the passengers.

He returned in about forty minutes and flew the rest of the passengers to the rodeo field. Though it was dry and solid, to Hunter it looked short for the big trimotor. Hunter was right. It was a bit short. On takeoff the plane's wheels brushed the tops of trees. Frank banked the plane with one wing low to clear other treetops.

By then Hunter was frightened, wondering what he had gotten himself into.

When, finally, they were headed for Prince George, Dorbandt reached under his seat, pulled out a bottle, swallowed a couple of gulps from it, turned and grinned at Hunter, crossed his fingers, and turned the airplane over to Don Glass and fell asleep.

Near Prince George, Frank made a fine landing on a not-very-smooth field. Hunter was *really* nervous by then, but didn't know what to do about it. It seemed he was stuck, for he did want to get to big game guide Andy Simon's place near Seward for a bear hunt by the scheduled departure date of May 25.

While the plane was refueled, the passengers ate at a nearby restaurant. The plane was next bound for Smithers, British Columbia, where there was fuel and an airfield. As soon as they were airborne, Frank again went to sleep after handing the controls over to Don Glass.

About 10 p.m. lights of Smithers showed up. Don woke Frank, who skillfully landed the overloaded trimotor on the small sod strip, aided by the lights of about fifteen automobiles. The drivers had heard the plane circling, and had driven out to light the runway to help with the landing.

None of this eased Hunter's nerves.

That evening Hunter made friends with a local citizen. They ate a meal together and had a good many drinks. The citizen consoled Hunter by saying, "It will be a great day for Smithers if you fellows can get off that little field in the morning with that big plane. Only one other ship has landed here before, and that was when Wiley Post stopped on his 1933 round-the-world flight. I sure hope you make it."

The next morning, May 23, workers cut a little opening in the jackpines at the end of the landing strip and shoved the tail of the trimotor into it to give Dorbandt the last foot of room. The takeoff from that short runway was without incident.

After passing Hazelton, the Ford left all signs of civilization. It flew over forest, lakes, and mountains, which extended east and west as far as the passengers could see. They passed to the west of enormous Babine Lake, and were soon 12,000 feet above a sea of sharp, cold and ominous mountains.

The next landing was on a primitive airstrip above the town of Telegraph Creek, where Frank had the plane refueled from five-gallon-tins. This took a long time, but Dorbandt put it to good use, retreating into the bushes to drink booze with a couple of old buddies from Telegraph Creek and Frank William's three companions.

Frank took off with only thirty minutes more fuel he estimated he needed for the flight to Atlin, British Columbia, 150 miles to the north. The big ship got off with no difficulty. Frank then startled his passengers by exuberantly executing a couple of wing-overs with the overloaded Ford for the benefit of his Telegraph Creek friends. He set a course for Atlin and told Don Glass to wake him when Atlin Lake appeared, and returned to his slumbers.

When a large lake appeared below, Don woke Frank, who appeared disturbed. He circled and found a settlement of a few cabins, and headed west. They had drifted east of course, and were over Teslin Lake, seventy or eighty miles from Atlin Lake with no chance of having enough fuel to arrive there.

When the fuel had nearly run out, Frank skillfully landed the ship on a high altitude short-grass meadow. They had landed between Teslin and Atlin Lakes, and between the small Sucker Lake and Surprise Lake, twenty-five or thirty miles from Atlin.

"It was mighty fine to step out of the ship onto solid ground," wrote Hunter.

Don Glass and Hunter walked to Atlin for help, arriving at about 6 a. m. the next morning, May 24.

This is where the accounts of Hunter and Frank Barr coincide. Hunter misnamed Barr, calling him Ralph Paar; otherwise describing his Moth biplane, and hauling gasoline to Dorbandt's downed Ford as Barr told it.

Hunter commented that, while he waited in Atlin for Dorbandt to arrive with the big Ford, those he talked with were outspoken in their dislike for Dorbandt because of his lack of participation in the unsuccessful search for Paddy Burke. This, of course, was also noted by Barr.

Shortly after noon, Dorbandt landed the tri-motor on the short Atlin strip, and no one was the worse for the experience. He had practically no fuel left, and made a straight-in approach to land, knowing he had no margin for circling.

At Atlin, Hunter took Dorbandt aside and accused him of being irresponsible, not fit to be in charge of a plane load of people, that he shouldn't be permitted to fly, etc. etc. He thought Frank might argue, and half-expected fists would fly. To his chagrin, Frank agreed with all he said. "Dorbandt burst into tears," Hunter wrote.

[Author's comment: Hunter's account of the flight seems to fit the known character of Frank Dorbandt except for this segment. It is difficult to believe that a dressing down by a young passenger would bring the arrogant, devil-may-care pilot to tears. That is, unless he had too generously partaken of John Barleycorn.]

Next, after landing and refueling at Whitehorse, the overloaded Ford headed for Dawson City, following down the Yukon River. There was no landing field at Dawson, but on May 25, Dorbandt successfully landed on a very rough, sloping hay field across the Klondike River from Dawson City. It was obvious the plane could not possibly get off from that field with its heavy load.

The passengers were taken to Dawson by boat. Fairbanks was now about 280 air miles distant.

Alone, Frank took off from the sidehill field and flew to a ranch meadow fifteen miles up the Klondike from Dawson, where two crawler tractors, a roller, and a grader were preparing an airplane runway. Freight and baggage from the plane was taken across the river by boat and trucked to the partly completed airfield.

After a four day wait, word came that Dorbandt would take a light load, consisting of Frank Williams and his people, on to Fairbanks, and return for the other passengers. Hunter watched the take off, made without difficulty, though the air strip was still very rough and soft from constant rains.Dorbandt was turned back by bad weather and snow; he had not reached Fairbanks.

This ended Hunter's account of the flight with Dorbandt. He fled Dorbandt and the trimotor, and with two prospectors he had met, floated down the Yukon River to Circle City, Alaska.

He rode a Model T Ford to Circle Hot Springs, and flew from there to Fairbanks, where he learned that Dorbandt had left for Anchorage two hours before his arrival.

Hunter finished his adventuresome trip by flying from Fairbanks to Anchorage with Frank Pollack, and chartering a float plane in Anchorage to take him to big game guide Andy Simon's Lakeview home near Seward.

ON ARRIVAL AT ANCHORAGE after his adventurous flight from Seattle with the Ford Trimotor, typically, Dorbandt announced himself by roaring over Fourth Avenue at 100 feet. He landed at Merrill Field, or what would become Merrill Field, and spent the next few hours selling hops over the city at $5 a seat.

Civil Aeronautics Administration inspector Murray Hall arrived in Anchorage later that day and heard about the new-in-town Ford Trimotor skimming low over down-town Anchorage. He hurried to Merrill Field to find the pilot of the offending aircraft.

Dorbandt, anxious to start operating from Anchorage, had not completed the paper work involved in setting up flight operations. Hall asked Dorbandt about his illegal low flight over Anchorage, and wanted to see his credentials.

"I'll show them to you later," he said. "Right now I'm busy making dollars."

"You're not flying anything for a while. And you're definitely not flying that Ford Trimotor. You and the airplane are both grounded," Hall pronounced. To verify, he borrowed a stepladder and wired red tags to all three of the Ford's props.[1]

In time Dorbandt established Ptarmigan Airlines at Anchorage; the Ford Trimotor he flew to Anchorage, the first of that breed to fly in Alaska, was dubbed the Ptarmigan II.

Though considered by other pilots as being too bold and wild, Dorbandt had a way with airplanes. Some of his exploits made him look pretty good to the Alaskan public.

Sometime during the early-1930s, Dorbandt flew a float-equipped Fairchild 71 to a cove on the Alaska Peninsula to pick up a badly injured commercial fisherman who needed immedi-

1. Don Glass arrived in Anchorage safely in 1934 on that flight with Dorbandt. Over many years he flew in Alaska for McGee, Star, and Woodley Airways. He died January 19, 1943, in the crash of a Woodley Airways Stinson A trimotor in Gastineau Channel near Juneau while trying to start a stalled third engine.

ate professional medical help. As he landed on a swell, one float hit the water so hard it bent that float's struts. Result; one of the Fairchild's wings drooped.

Dorbandt taxied to the fishing boat, and with help from the crew heated the strut with their blowtorch and managed to straighten it as the boat and airplane bobbed about in the swells. He loaded the injured fisherman, taxied through the swells, took off, and flew the man to Anchorage.

He even made a return flight to Siberia, less than a year after he had flown there with Eielson. A. P. Jochimsen, captain of the Olaf Swenson-owned *Karise*, which had replaced the ill-fated *Elisif,* was seriously ill. The *Karise* radioed for a plane. Dorbandt, with his mechanic Lon Cope, responded with the Fairchild 51 NC5364 leaving Anchorage in the late afternoon of September 20, 1930.

Dorbandt and Cope spent the night at Unalakleet, flew to Nome to refuel next day, made the flight to Cape Serdtse-Kamen in Siberia, landed on slushy water, and taxied to the *Karise* to pick up Jochimsen.

They refueled and immediately flew to Teller where they spent the night. Next day they delivered the seriously ill captain to the Nome hospital.

Frank Barr wrote the following about another of Dorbandt's exploits. Arriving over the Anchorage airport with a load of illegal fur and the buyer who owned it, Dorbandt spotted the car of Alaska Game Commission Wildlife Agent Jack O'Conner.

"I don't dare land here," he told the fur buyer.

"What will you do? What about my fur?" the buyer asked.

"I don't have enough gas to go anywhere else. We'll have to fly a few miles out of town and shove the fur out the door. We'll come back and land empty. O'Connor won't know the difference."

"I've got thousands of dollars tied up in that fur. I don't want to lose it," the buyer complained.

"I've got plenty tied up in this airplane, and I don't want to lose it *and* go to jail,"Dorbandt responded. "Just calm down. I'll find a place close to the road where we can drive a car and pick up the fur," he told the buyer.

There were probably genuine tears in the fur buyer's eyes as he shoved his precious commodity out of the circling plane. Dumping the fur probably had no effect on Dorbandt. He expected to be paid for the charter, regardless.

It took a couple of days to locate and retrieve all the fur. Of course, Dorbandt probably exaggerated the details as he told and re-told the story.

Dorbandt was always willing to fly anywhere and do about anything for money. He made many flights that other bush pilots turned down. One of these was one made for Father Bernard J. Hubbard, head of the Geology Department of California's Santa Clara University, who was sometimes called "The Glacier Priest." Hubbard chartered him to make a landing on the two-mile-long lake in the caldera of Aniakchak volcano, which lies on the Alaska Peninsula, 200 miles southwest of Lake Iliamna.

Dorbandt popped his float plane over the rim of the crater, landed on the lake, and let Hubbard and his assistant off. While he and his mechanic, Herb Larison, waited for their passengers to complete their field work, they checked their fuel supply. It was much less than expected. They had been forced to fight weather after refueling at a cannery.

Dorbandt warned Hubbard, "We're low on fuel. The nearest fuel is at the cannery, about twenty minutes flying time if we don't have to dodge weather. We could make that fine, but it's going to take an extra ten minutes to climb out of this hole. I'm not sure we'll have enough gas to get to the cannery."

Dorbandt took off, flew to the rim of the crater, and in three or four minutes found an upward flow of wind to help carry him up and over the edge. He made it to the cannery bay, started to taxi ashore, and ran out of gas. A fishing boat towed the plane ashore.

Hubbard had approached several other bush pilots about taking the charter, but none were interested. Any flight was sufficiently hazardous in that remote region without a descent into the caldera, where winds could be unpredictable and fumes might be hazardous. But Dorbandt probably didn't give it a thought; he was arrogant, and believed his skill could pull him through just about anything.

Frank Dorbandt's luck ran out at a Yukon River village, where he spent a night. Next morning, while pulling his propeller through to start the engine, he slipped on the ice and the spinning prop hit and nearly severed one of his hands at the wrist.

He waited three days for another plane to land to take him to medical care, but none arrived. He then, one-handed, flew himself to Fairbanks. After treatment, he remained at St. Joseph Hospital for several days. He gave the nurses and sisters such a rough time with his ribald sense of humor and crude jokes that he was discharged at the earliest possible time.

He celebrated his release from the hospital by visiting a few of the local bars. A day or so later, an acquaintance found him ill at his hotel. He returned to the hospital, where he died a day later of pneumonia.

The account of his death in the *Daily News-Miner*, instead of the usual eulogy, said simply, "After arriving in Alaska, Frank Dorbandt carved a unique career for himself."

Dorbandt first flew in Alaska in 1929 for Anchorage Air Transport; he later flew for Wien Alaska Airways, Alaskan Airways, Dorbandt and Cope Flying Service (later called Pacific International Airways), and Ptarmigan Airlines.

Word of his rough challenge to Ben Eielson, which some believed caused Eielson to make his fatal flight, clung to Dorbandt. Many blamed him for Eielson's death. Perhaps he did precipitate Eielson's last flight. However, an airplane pilot, like a ship's captain, is responsible for all decisions made relating to his ship. To decide to go, or not to go, is the pilot's responsibility. Eielson may have been challenged by Dorbandt, but in the end he alone decided to make his ill-fated last flight.

Floyd Andrew DeHayes, alias Frank Dorbandt, left his mark as a capable, bold pilot, who, along with others, pioneered Alaska's skies from his arrival in the Territory in 1929, until his death. Unfortunately, his blustering, crude, challenging, and unpleasant demeanor detracted from his accomplishments.

He is memorialized by Dorbandt Street, in Anchorage.

19 Noel Wien; the Later Years

Noel Wien could have remained a pilot for Alaskan Airways, Inc. which in 1929 bought Wien Alaska Airways, the company he and his brother Ralph had built. He chose not to, and with the sale, agreed that he would not fly for any competing company, or start his own company in Alaska, for three years.

He and his wife Ada traveled Outside, where Noel looked into the possibility of a flying job. He was now famous, hailed as "The Lindbergh of the North," and reporters followed the couple wherever they went. Within a couple of years articles were written about his aviation pioneering in Alaska in at least fifteen magazines, and dozens of newspapers and their Sunday supplements.

At Wayne, Michigan, in the fall of 1929, Noel bought a new Stinson Junior, NC490H with a Wright J5 220-hp air-cooled engine. The couple visited the Wien family at Cook, Minnesota, and Noel barnstormed with the new Stinson, appearing with it at various state and local fairs.

News that Ben Eielson was missing reached Noel in November. He wired Alaskan Airways at Fairbanks offering to help search for the Hamilton, but word came back, "Thanks, but no need. There are enough pilots now."

Though he wasn't present on the Eielson search, his and Ralph's winter experience of flying and maintaining aircraft throughout winters, as well as the same experience of the Rodebaugh-Bennett Company, clearly was helpful to the pilots and mechanics who searched for the Hamilton. Flying and keeping airplanes running in the deep cold is an art as well as a complex mechanical challenge. Mechanics and pilots in Alaska commonly transferred

The Wien Ford Trimotor that Noel Wien bought from Northwest Airlines with payment for his flight Fairbanks-Seattle with photos of the Wiley Post/Will Rogers crash. Here it is on a tiny air strip at Jack Wade, in the Forty-mile country. Horses were far more common than airplanes in interior Alaska during the 1930s.

between companies, and the know-how to keep 'em flying in the deep cold of winter was general in Alaska's fledgling commercial aviation industry.

Job opportunities existed for pilots outside of Alaska, but Noel didn't commit himself to any of them. One of the most likely offers was for him to become Chief Pilot on a Saint Paul-to-Winnipeg route planned by Northwest Airways. He waited months for the position to firm up.

He decided to return to Fairbanks to accept the job offer to fly for Alaskan Airways. His wife Ada, and baby Merrill,[1] traveled by train to Seattle, by ship to Seward, and by train to Fairbanks.

In December, 1930, Noel, with his younger brother Sig, a master mechanic, flew the Stinson Jr. from Minnesota to Fairbanks.

Soon after arriving in Fairbanks, Noel had many job offers. One

1. Born April 4, 1930, in Minnesota, he was named for Russell Merrill, pioneering Anchorage bush pilot who had become a close friend of Noel's. Merrill went missing on a flight at the time Noel and Ada left Alaska in September, 1929. He was never found. Merrill Field, at Anchorage is named for him.

was from Charles "Speed" Holman, Northwest's Chief Pilot, with a firm starting date for the Northwest Airlines position for which he had waited months. He turned Holman down. Like so many who have lived in Alaska and left only to return, this huge, mostly unsettled land had a firm hold on him. Now it felt like home.

Noel sold Alaskan Airways the Stinson Jr., and, until his noncompetitive agreement ended in August, 1932, he flew from Fairbanks as one of the company's regular pilots. With the end of the agreement, he started Wien Airways of Alaska out of Fairbanks, Nome, and Anchorage with a six-place Bellanca Pacemaker NC345W powered with a 300 hp Wasp Junior.

Noel continued to fly year-round. He was wary of deep cold flights, but occasionally made them. On one occasion, with the Bellanca, he flew 1,500 pounds of groceries to Wiseman when the temperature was -40 F.

On another occasion, he allowed himself to be talked into flying a passenger in his Bellanca from McGrath to Fairbanks when the plane's airborne engine temperature stabilized at 100 degrees, far below normal. The ground temperature at McGrath was 65 below zero; at Fairbanks it was 66 below. His emergency gear was not sufficient to maintain his and his passenger's life on the ground. If he had been forced down, both would probably have quickly frozen to death.

He couldn't land at the Fairbanks airport because it was hidden by dense ice fog. Instead, he landed in a small clearing just outside of town. His passenger suffered a frostbitten nose while rushing from the airplane to a nearby farmhouse.

He had hung up another record; by flying an airplane on the coldest day ever recorded at Fairbanks.

His comment? "If I continued to operate in this manner, I was not long for this earth."

Survival in Deep Cold

Noel Wien wrote the following about a cold weather trip in an issue of the Arctic Liner, *the Wien Company magazine. It was reprinted in a booklet published in 1967 by Wien Air Alaska called* The Wien Brothers Story, *by Kay J. Kennedy.*

"Midwinter of 1933 I was flying freight from Fairbanks to Beaver, and from Beaver into the Chandalar area. After two trips from Fairbanks I had four to make from Beaver to the mining operation at Big Squaw Lake with supplies purchased from the Beaver trader. The daytime temperature was near minus -40 F., and daylight was not more than six hours duration. I planned to make three trips to Big Squaw in one day, the fourth trip the following day, and then return to Fairbanks.

"On the return flight of the third trip it was almost dark so I decided to land at nine-mile-long Chandalar Lake, about twenty miles from Big Squaw, slightly off course to Beaver. I had seen an old cabin there and thought I could be quite comfortable for the night. The wind blows the snow in hard drifts on this long lake, which lies north and south.

"The cabin was in a little cove of the lake, and fortunately, the snow was a little less rough there than on other areas of the lake. I could not see the drifts in the darkness. Although the landing was extremely rough, the landing gear was not damaged.

"It took about an hour to cut spruce boughs to put under the skis and to drain the oil. I then wallowed through the deep snow to the cabin. A quick glance told me I would not be as comfortable as I had hoped. Most of the chinking between the logs of the walls and roof of the one-room-cabin was gone. About half of the five-gallon can used for a stove was burned out. In one corner of the room was a bunk built from poles—no springs, mattress or blankets. Otherwise the room was bare. There was no door on the cabin, no glass in the window.

"Luckily I had to go only a few feet from the cabin to cut firewood. It was hard to get the fire started and to get it to draw through the cold, crooked, rusty, full-of-holes, stove pipe. Once I had a good fire going, I decided not to attempt to cook anything to eat, but rather to crawl into my wool sleeping bag stretched over the pole bunk.

"I soon found the poles were most uncomfortable, the sleeping bag inadequate, and that I had to feed wood to the fire constantly in order to keep even slightly warm. I did wear good winter clothing—fur mukluks, caribou skin pants, and a long fur parka. The rest of the night I sat in front of the stove with the sleeping bag draped over my shoulders and out on the sides to deflect the heat around me.

"As I sat through that long night, frequently putting wood into the little makeshift stove, I had plenty of time to think. In my previous ten years of flying I had learned many lessons the hard way. Not until this night did I realize how ill-equipped I was to survive should I have to camp out in severe temperatures for one or many nights.

"First, I decided I would carry nothing but the best Four Star Woods eider down sleeping robe, or one made of caribou fur turned inside out. No more wool blanket bags for me. I also decided I would have a small tent made from light canvas or airplane cloth, and my own five-gallon gas can stove with three lengths of small pipe. The pipe and tent could fit inside the stove, take little weight or room in the plane, and be readily set up in case it should be necessary to camp out overnight. How much more comfortable I would have been on this night if I had such equipment with me!

"At daybreak the following morning it was comparatively easy to get the plane warm and airborne. I had the necessary firepot and motor cover with me. This incident occurred long before plane and ground radio, and weather facilities were a reality in Alaska. The folks back home, not expecting me for a few days, did not know if I was in trouble. No search plane, if available, would be sent to look for me until considerable time elapsed.

"Many of the do's and don'ts of Arctic flying which seem so obvious today had to be learned by trial and error. One did have the advantage of being able to learn from the experiences of others. Many such incidents, so much more serious than this one, made me realize what serious trouble I could get into without proper emergency equipment."

The First Commercial Passenger Flight Fairbanks to Seattle
By Noel Wien

Another first flight recorded by Noel Wien occurred when he made the outstanding flight of the year in response to a hurry-up call to fly hot news photos from Alaska to Seattle. He wrote about it in a Wien Arctic Liner *and Kay Kennedy included it in her 1967 booklet,* The Wien Brothers' Story, *as follows:*

"On August 15, 1935, internationally famous, admired, and loved Wiley Post and Will Rogers were killed in the crash of their plane near Barrow, Alaska. Immediately after the first tragic news was received, two Alaska air carriers were hired to rush to

Seattle pictures to be taken at the scene of the accident, and the last pictures available of the two men in Fairbanks prior to their departure for Barrow. Wien Alaska was one of the hired carriers.

"Chester Brown in a Wright powered Bellanca on floats carried Alfred Lomen and photographer Emil Jacobs from Nome, via Kotzebue, to the lagoon fourteen miles south of Barrow, the scene of the crash. There Jacobs hurriedly took pictures and the three men then took off for Barrow to gas up for the flight to Fairbanks. However they were unable to obtain aviation gas there, so took on enough low grade gas and flew back to Kotzebue. There they were able to obtain aviation gas and proceed to Fairbanks.

"Although northern Alaska still had no dark nights this time of year, most of the trip was made through extremely adverse weather—low fog and rain. The plane had no radio and weather reports were not available from points in between. When Chester arrived in Fairbanks, he was exhausted.

"Thirty minutes after their touchdown arrival in Fairbanks, Victor Ross and I took off in a Wasp, Jr. Bellanca on wheels with Alfred Lomen as passenger. This Bellanca carried 112 gallons of gasoline. In the cabin we loaded twelve five-gallon cans of gas. The plane was equipped with a wobble pump so that, between refueling stops, passengers and pilot working together, could pump gas from the cans to the plane's tanks.

Night Flight

"We took off from Fairbanks at 2:30 in the afternoon. The weather was clear all the way to Whitehorse. We landed there four hours and fifty-five minutes later. At Whitehorse we cleared customs, took on more gas in the tanks and more cans of gas in the cabin. We remained there until 11 o'clock in the evening so that we could make a daylight landing at Prince George the following morning. The flight from Whitehorse to Prince George was 650 miles over a mountain range seven to eight thousand feet high. Although the weather was clear, it was dark, no moonlight, with only the stars and the silhouetted outline of the mountains to guide us. At intervals during the night Alfred Lomen opened gas cans, put the hose into the cans one at a time, and I pumped the gas from the cans into the tank.

"We landed at Prince George at six in the morning, lost two hours first trying to find a customs officer and then clearing customs for Seattle. After refueling we took off for Seattle, and arrived four hours and fifteen minutes later.

"We were met at the airport by representatives of the International News Service who literally grabbed the film out of our hands. Two hours later when the rival plane landed in Seattle, newspapers carrying the pictures and story which we had brought to Seattle, were on the news stands.

"At the end of the year, the Post/Rogers story was listed as the second largest newspaper scoop of 1935."

[AUTHOR: *This flight by Noel Wien has long been considered one of the most remarkable of the period. It was the first airplane to follow the inland route Fairbanks to Seattle. The seven-hour moonless night flight across the high peaks of the jagged Cassiars between Whitehorse and Prince George in particular was an amazing accomplishment. Loss of an engine at any point could easily have been fatal. A few degrees off in navigation over the 650 miles, or being pushed off course by an unknown crosswind, could have found the airplane low on gas and miles from the goal of Prince George. However neither happened. That Noel accurately landed the plane at Prince George in a timely fashion was a truly great feat of navigation.]*

THE FORD TRIMOTOR

Noel received $3,500 for the flight [Fairbanks to Seattle, with photos], based on $50 per flying hour for the round trip. While in Seattle, he used this money to make a down payment on a Northwest Airlines surplus Ford Trimotor.

This all-metal airplane, NC8419, had three 420-horsepower Pratt and Whitney Wasp engines, and a fuel capacity of 555 gallons. It had the first voice radio in any plane in Alaska, carried twelve passengers, had so-called air wheels—sixteen-inch-wide balloon tires that supported it on soft ground, and a load capacity of 3,300-pounds. This was far under what the "Alaska load capacity" could be. It had disk brakes, important in Alaska with its many short fields; it could stop quickly after landing.

The price of this workhorse of an airplane was $7,500, a bargain even in that year of 1935.

The Harvey Goodale portrait of Noel Wien, with his Ford Trimotor in the background. Courtesy of the Bob Reeve family.

Noel was checked out in the huge (for Noel, and for the time) three-engine airplane by a Northwest Airlines pilot in a mere hour and ten minutes before he prepared to leave for Alaska with it. On August 28, after waiting for days for fog to clear, Noel and Ross took off with the Ford for Fairbanks with several passengers, including Alfred and Ralph Lomen.

First stop on the return flight was at Prince George, 475 miles north of Seattle where they remained overnight. The next day the weather was fair, but low ceilings farther north necessitated a landing at Hazelton, 200 miles northwest. The field at Hazelton was Noel's first test in putting the 14,000-pound-gross plane into a small field; this one was only 1,500-feet long.

The approach had to be made over the river to the field because there was a fence at the other end, and steep bluffs three to four hundred yards beyond. The take-off also had to be made over the river. Flying the big Ford came easy to Wien, and the landing and takeoff went well.

The remainder of the flight to Fairbanks was made without unusual incident.

Noel shortly made three round-trip passenger flights to Seattle with the Ford, and though there was demand for the service, flights to Seattle were discontinued. There were few landing fields on the route, a lack of weather reports, no ground facilities, no radio communication. Each trip required being away from Fairbanks for one to two weeks.

There was not enough income for the time involved. There was plenty of passenger and freight business for the Ford between Fairbanks and Nome to keep the big bird busy.

A spectacular start of one of the flights to Seattle with the Ford Tri-motor took place in May, 1936, at the Bessie Creek runway at Nome. There were ten passengers, lots of baggage, and full tanks of gas. The big plane was quite heavy. Snow was mostly cleared from the runway, and a puddle of half-frozen water decorated the strip. Noel thought he could run through the water with the right wheel on takeoff without a problem.

He was wrong.

Airplanes were parked on each side of the first part of the runway, and while running between them, he was careful to use about three-quarters power; he planned to use full power after he was clear.

Noel Wien (right) and Sam O. White, a former federal Wildlife Agent in Alaska, later a much respected bush pilot who flew for Wien Airlines for many years. The airplane behind White is a Fairchild 71; the one behind Wien is probably a Stinson. Circa 1930s.

When the big Ford's right wheel hit the puddle, the plane abruptly veered to the right toward the edge of the runway. Noel immediately cut the power from the left engine, and slammed the throttle to the right engine to full power. The right engine didn't respond; it simply idled.

Now he had full power on the center engine, half power on the left, and virtually no power on the right. The plane was moving fast, and he couldn't take his hand off the wheel to pump or jiggle the throttle for the right engine to bring it to full.

At nearly takeoff speed, the Ford left the runway and ran into deep snow, circling right. By then it was almost flying, and the wheels left tracks in the top of the drifted snow.

The right engine, though set for full power, continued to idle daintily.

The half-power left engine, and the full power center engine kept the airplane partially airborne, with the wheels lightly running on top of the snow. The off-center power took the airplane on a 1,500-foot run in a 120-degree circle. Noel had to hold the control wheel back to keep the tires from breaking through the deep snow, and he couldn't take his hand off of it to coax the right engine to full power.

Finally, on its own, the right engine decided to come alive with full power. Noel then brought the left engine to full power, and the Ford nicely lifted off the snow. He then pointed the three roaring Pratt & Whitney Wasps toward Seattle.

Building an Airline

Noel was the pioneer. Brothers Sigurd and Fritz joined him, and the three built Wien Airways, absorbing several other bush airlines, and merging with Northern Consolidated Airlines, finally becoming Wien Air Alaska. By 1963 the airline had 4,500 miles of certified mail routes. Until 1985 the company was a major airline that provided jet air service in Alaska, as well as in the western South 48 states. Economics, union problems, and mismanagement (not by the Wien family) resulted in the dissolution of Wien Air Alaska.

Noel Wien was quiet and unassuming. No early pilot had a finer reputation. "Any commitment Noel Wien made was like a gold bond," fellow pioneer pilot Bob Reeve once said.

Noel was the recipient of many honors; in 1962 the University of Alaska Fairbanks awarded him an honorary doctor of science degree; to commemorate the golden anniversary of his 1924 first flight Anchorage-to- Fairbanks, Governor William Egan declared a Noel Wien Day in Alaska on July 15, 1974; Noel was Alaskan of the Year 1975; the Fairbanks Public Library, built on the site of Fairbanks' onetime Weeks Field (which closed in 1951), is now the Noel Wien Memorial Library.

Wien Mountain in the Arctic Brooks Range, 6,000 feet high, two miles north of Mt. Doonerak and thirty-seven miles northwest of Wiseman, was named about 1930 by Robert Marshall after Noel Wien, first aviator to land in the Koyukuk country, and the first to fly over this peak.

There is Wien Street, in Fairbanks, which connects 2nd Avenue and Cowles Streets.

Wien Lake, four and a half miles long, twenty-two miles southeast of Bitzshini Mountains, in the Kilbuck-Kuskokwim Mountains, was locally named for Noel Wien who once landed there with a float plane. His long-time friend and fellow pilot, Sam O. White, recommended this naming to the U.S. Geological Survey.

Noel was inducted into the Minnesota Aviation Hall of Fame in 1989, and into the Alaska Aviation Heritage Museum's Hall of Fame in 2000.

Noel Wien died on July 18, 1977. He was 78. Immediate family survivors were his wife Ada, sons Merrill and Richard, and daughter Jean.

20 Other Participants in the Eielson/Borland Saga

NEWSPAPERS AROUND THE world wrote exciting copy about the ice-locked *Elisif, Nanuk*, and their "million dollar" cargos of fur. For weeks names of the pilots who flew in search of the lost Hamilton, and who flew fur back to Alaska, were headlined. Little appeared in print about others who played a role in this arctic drama, including, for example, Robert J. Gleason, Herbert Larison, and James Hutchison.

ROBERT J. GLEASON

Gleason was the 23-year-old radioman on the *Nanuk* through that winter of 1929-30, and with his radios he was the center of intelligence and communication for both the search for the lost Hamilton, and the aerial transportation of fur.

He was born in Seattle, Washington, in 1906, and grew up there. While in the Boy Scouts, he and his boyhood friend Charles Huntley became interested in telegraphy, and both earned their commercial telegrapher's licenses at an early age. In 1929, Huntley was the radioman on the *Ilisif,* while Gleason was the same on the *Nanuk.*

When Gleason sailed on the *Nanuk* he had completed his junior year studying electrical engineering at the University of Washington. After his time on the *Nanuk,* he returned to the University to graduate with a degree in electrical engineering.

Before his service on the *Nanuk*, he had worked three springs and summers in Alaska as a radio operator, first with the U.S.

Aircraft mechanics at Fairbanks who worked for Pacific Alaska Airlines (a branch or Pan American World Airways) in 1934-35. From left are Ed Moore, Warren Tilman (who flew as a mechanician with Ed Young), Jim Hutchison, Orval Porter (who accompanied Sir Hubert Wilkins, Eielson, and Crosson to Antarctica), Gordon Springbett, and Austin Gibbs.

Corps of Engineers in Wrangell Narrows, and in 1927 and 1928 at the salmon cannery at Kake, both in Southeastern Alaska.

His contact with Alaskan aviators while he was the radioman on the *Nanuk* led to his lifetime profession of aviation communication. From 1932 to 1942 he was Chief Radio Operator and Engineer for Pan American Airways at Fairbanks, where he was responsible for the construction and operation of the first airline radio system in the Territory. He installed PAA radiotelegraph equipment in the Fairchild 71 airplanes, and other planes the company then flew, as well as building the necessary ground radio stations.

In August, 1934, he returned to Siberia, flown from Fairbanks to the coastal village of Uelen by Joe Crosson. In the spring, 1974, issue of *The Alaska Journal* he wrote, "I had spent the winter of 1929–30 at North Cape, Siberia, frozen in on the schooner *Nanuk*. That was where I first met Joe. Crosson and I were back together again!

They had met, of course, during the Eielson/Borland search. Pan American Airways was interested in developing flights through Alaska, across the Bering Sea, through Siberia and the Orient. They had authorization from the U.S.S.R. to make an initial flying survey from Nome to Uelen, Siberia, and on down the Siberian seacoast to Anadyr.

Included on the flight was Harlee Branch, Assistant Postmaster General of the U.S., Robert Thach, senior Pan Am Vice-President, and Lyman Peck, General Manager of PAA in Alaska. Gleason was Crosson's radioman on the flight. Crosson was PAA's Alaska Operations manager and was to appraise Russian airfields and facilitites.

The survey was a bust; the Russian Governor at Uelen refused to let it proceed because he had not been informed of it by his higher-ups. After waiting at Uelen for three days for permission to continue, Crosson flew the party back to Fairbanks.

Pan Am lost interest in the Siberian route to the Orient. Instead, the company developed the Pacific route. In October, 1936, it started flying huge Martin M-130s (China Clipper flying boats) from San Francisco to seaplane bases at Honolulu, Midway, Wake, Guam, Manila, and Hong Kong.

During World War II, from 1942 to 1944, as a Lieutenant Colonel in the U.S. Air Force, Gleason was in charge of all Air Force communication in Alaska. Then, through 1945, he became Group Commander, USAF, in India and China, for which he received the bronze star.

Returning to Pan American Airways after WWII, Gleason served again in the Pacific-Alaska Division, then in the Latin American Division until 1949, when he joined Aeronautical Radio Inc. This firm provided nationwide communications service to all airlines. He was executive vice-president and director of the firm from 1955 until his retirement in 1971.

His book, *Icebound in the Siberian Arctic*, describing his winter of 1929-30 aboard the *Nanuk*, was published in 1977 by Alaska Northwest Publishing Company. He dedicated the book to his friend Joe Crosson.

HERBERT M. LARISON

On April 3, 1930, Joe Crosson and Ed Young, both flying Fairchild 71s, flew from Teller to the *Nanuk*. With Crosson was Herbert M. Larison, a mechanic who worked for Alaskan Airways at Fairbanks. Larison's assignment was to start repairs on the Fairchild 71 CF-AJK that Pat Reid had broken on February 23 when he landed with a load of gas at the Hamilton wreck site. Larison had with him two sacked prop blades, one ski stand, and two replacement skis for CF-AJK.

Crosson left Larison at the *Nanuk*. The mechanic then traveled by dog team the ninety miles to the wrecked Fairchild, which lay a few hundred feet from the remains of the wrecked Hamilton. There he dug a cave in the four-feet-deep snow beneath the bellied-in plane, and in ten days of solitary work, while camping out, did all the repair work possible.

The nearest human was trapper Brokhanov, whose cabin was six miles distant. It must have seemed a lonely world to Larison as he toiled away, in a desolate, snow-buried foreign country, hours flying time from Alaska. The sad reminder of the nearby crashed Hamilton probably didn't improve his mood.

Welds were needed to complete repairs. By dog team he returned to the *Nanuk* where Bob Gleason radioed Fairbanks, asking that master welder Jim Hutchison and his equipment be flown to the downed plane.

On April 27, Crosson with pilot S.E. Robbins aboard, flew Hutch and his welding tanks to the downed Fairchild 71 CF-AJK. With the welding completed, Larison and Hutch shoveled a snow ramp in front of the repaired plane, and Crosson taxied it up and out, and flew it to the *Nanuk.*

Robbins followed with the other Fairchild. At the *Nanuk* the two Fairchilds were loaded with forty bales of fur, which they flew to Teller.

Finally, on May 8, Crosson flew a Fairchild 71, with pilot Robbins aboard, to the *Nanuk.* They remained overnight, and next day, with Crosson flying the Fairchild and Robbins flying the Waco 10 NC180C, (the "discovery" biplane Crosson flew when he and Gillam found the Hamilton), which had remained at the *Nanuk* for months, they returned to Teller. Larison rode with Robbins, his lonely five-week assignment in Siberia completed. The two ships carried twenty-six bales of fur. Now all Alaskan Airways flying equipment was back in Alaska, except for the remains of the Hamilton, which Soviet historians would return to Fairbanks some six decades later.

James T. Hutchison

Born in Pennsylvania in 1900, James T. Hutchison arrived in Alaska in 1919 with the U.S. Army. He arrived in Fairbanks in 1922, and was one of the six mechanics working at Fairbanks when Alaska Airways was formed in September, 1929. The new company owned the former Wien hangar at Fairbanks and ten airplanes, with two more planes on order.

Hutch, as he was called, knew and worked on the airplanes of virtually all of the early pilots who flew from Fairbanks.

He took flying lessons from Ralph Wien, but when Ralph died in a crash at Kotzebue in 1930, he decided to stick with being a mechanic, and abandoned the idea of becoming a pilot.

In his time as an aircraft mechanic, Hutch participated in virtually every important aviation epic that occurred at or near Fairbanks.

Hutch worked on aircraft for the Detroit News Expedition led by explorer/navigator Hubert Wilkins in 1926 and 1927. In 1931 he repaired damage to Wiley Post's *Winnie Mae* on Post's first around-the-world flight.

Jim Hutchison (Hutch), veteran aircraft mechanic who was involved with aviation at Fairbanks from the very beginning. He participated in many spectacular aviation events, including being flown to Siberia to repair a damaged Fairchild 71 that was involved in the Eielson/Borland search. Fairbanks Daily News-Miner

He also built a special heater for Joe Crosson's Super Swallow, in which Joe made the first commercial flight from Fairbanks to Barrow.

In the early years, Hutch often flew as a mechanic with bush pilots, in winter especially, when care of planes was labor intensive, and mechanical problems were fairly common.

The Federal Aviation Administration named Hutchison Alaska's Mechanic of the year 1972 for a design change he made on the nose wheel steering mechanism of Hercules aircraft; twenty years later the agency awarded him a "Master Mechanic" award for his lifetime of excellence in maintaining aircraft.

The $23 million Hutchison Career Center at Fairbanks, which includes a stand-alone career and technical high school, and a wide variety of certificate/associate programs, including an acclaimed airframe and power mechanics aviation program, is named in his honor.

Hutch and his wife Helen raised ten children in their log home in Fairbanks. Hutch died in Fairbanks at age 95 on September 9, 1995.

While pilots often make headlines with their sometimes spectacular and newsworthy exploits, those headlines would not be possible without mechanics like Hutch, Earl Borland, Herb Larison, and their fellows who keep the airplanes these pilots fly airworthy.

The *Nanuk*

In 1932 the *Nanuk* was chartered by movie maker Metro-Goldwyn-Mayer. That summer Captain Carl M. Hansen ran her to Teller, Alaska, where that fall she was allowed to be frozen in the arctic ice to be used in the feature movie *Eskimo; Mala the Magnificent*. Aboard was movie equipment, a movie-making crew, and actors. Director William S. Van Dyke was in charge of the filming, and Peter Freuchen, who wrote the book on which the screenplay was based, was also aboard. In the movie, Freuchen played the role of the captain of the *Nanuk.* Anna May Wong assumed the role of the Eskimo heroine.

MGM bought the *Nanuk* in 1933, renamed and remade her into the *Hispaniola* for the movie *TreasureIsland*. After that she became the *Pandora,* for the film *Mutiny on the Bounty*.

In 1941 she was reported to be laid up ashore at Long Beach, California, where she probably rotted away, unknown and unsung, like so many other stout old schooners of her time.

References

Periodicals

Barr, Frank, "One of a Kind," (career of Frank Dorbandt), *Alaska Magazine*, May, 1972.

Canadian Encyclopedia (Internet), "Where in Time is Sir Hubert," January 7, 2009.

Ellis, Bob, "The Mattern Expedition," *Air Alaska*, November, 1984.

Flight North, 1934 (ms.) Author unnamed. 23pp typewritten account of flight from Seattle to Alaska with Frank Dorbandt in Ford Trimotor.

Gleason, Robert J., "Recalling the '30s Flights to Siberia," *Air Alaska*, May,1989.

Gleason, Robert J., "Pioneer Mail Flight to Siberia," *The Alaska Journal*, Spring, 1974.

Jarmin, Lloyd, "The (in) Famous Fairchild '71—Ugly, Dependable, " *Air Alaska*, December, 1985.

Jarman, Lloyd, "If the Lockheed Vega was Being Built today it'd be Competitive!" *Air Alaska*, March, 1985.

Jarman, Lloyd, "The Vegas were Tough Birds, but Plywood Rotted in Sea Water," *Air Alaska*, March, 1985.

Jarman, Lloyd, "Dogfight Over Icy Strait," *Air Alaska*, July, 1987.

LeCompte, Tom, "The Disorient Express," *Air & Space, Smithsonian*, September 2008. [on blind flying]

Long, Everett, "New Evidence Sparks Interest in N-209," *Air Alaska*, October, 1989.

Long, Everett," Levanevsky Recalled, April Search Set," *Air Alaska*, February 1990.

Lincoln, Nicholas Sr., of Kenny Lake, Alaska. Letter to author with recollections of Harold Gillam, December 13, 1979.

Ohio State University Register of Sir George Hubert Wilkins Papers, Jan. 7, 2009 (Internet).

Perrigo, Dalene, "Glass crashes in Gastineau Canal," *Anchorage Times*, October 29, 1983.

Ruotsala, Jim, "Carl Ben Eielson, Father of Alaskan Aviation; The Early Years, 1922–24" *Air Alaska*, November, 1982.
Ruotsala, Jim, "Crash Marks End of Eielson Era," *Air Alaska*, April, 1984.
Ruotsala, Jim, "The 71-A Workhorse," *Air Alaska*, April, 1987.
Ruotsala, Jim, "Eielson and Wilkins were First to Conquer the North Pole via Plane," *Air Alaska*, June, 1989.
Ruotsala, Jim, "Joe Crosson: Working Pilot in the Best of Alaska Traditions," *Air Alaska*, June, 1988.
Spencer, Ted, "First Mail Flight to Siberia," *Wings Over Alaska*, Feb. 3 (No. 1), 1983.
Spencer, Ted, "Stearman is Valuable Surviving Historical Alaskan Aircraft," *Air Alaska*, November, 1986
Tilman, Warren O., Letter, 18pp, single-spaced, typed, addressed to "Clyde and all" (Tilman's sons). April 3, 1981. Copy to Author Rearden. Subject: Ed Young, pilot.
Trim Tab, "Ptarmigan Airlines," Issue 31, May 1, 1972.
Wachel, Pat, "Harold Gillam, an Alaska Fearless Flier," *Northern Lights*, undated clipping.
Wien, Noel, *Wien Arctic Liner* (Wien inhouse magazine). August, 1956; September, 1956; October, 1956; January, 1957; April–June, 1957; Fall, 1961.
Wings Over Alaska, Alaska Aviation History News. Beyond Blue Skies. [James Hutchison obituary], Vol. 3, Number 1, 1997.

Books

Alaska Almanac, The, 2006. Alaska Northwest Books, Seattle, Anchorage.
Allen, Frederick Lewis, "Lindbergh Flies the Atlantic," *The American Reader*, 1958. Rand McNally & Company.
Billberg, Rudy, as told to Jim Rearden, *In the Shadow of Eagles*, 1992. From Barnstormer to Alaska Bush Pilot, A Flyer's Story. Alaska Northwest Books, Anchorage.
Brooks, James W., *North to Wolf Country; My Life among the Creatures of Alaska*, 2003. Epicenter Press.
Bush Pilots, The; Time, 1983, Time Incorporated, Chicago.
Chronicle of Aviation, 1992. J.L. International Publishing, Liberty, Missouri.
Doolittle, James H., with Carroll V. Glines. *I Could Never Be So Lucky Again.*1992. Bantam Books. [on instrument flying]
Encyclopaedia Britannica, 1961. Vols. 1, 2, 4, 7, 16, 20, 21, 23.
Gleason, Robert J., *Icebound in the Siberian Arctic.* 1977. Alaska Northwest Publishing Company, Anchorage.

Griese, Arnold, *Bush Pilot; Early Alaska Aviator Harold Gillam, Sr., Lucky or Legend?* 2005. Publication Consultants, Anchorage.

Green, William, and Pollinger, Gerald, *The Aircraft of the World*, 1954. Hanover House, Garden City, New York.

Harkey, Ira, *Pioneer Bush Pilot; the Story of Noel Wien*. 1974. University of Washington Press. Seattle.

Kennedy, Kay J., *The Wien Brothers' Story*, 1967. *Wien Air Alaska*, Fairbanks (booklet).

McKinley, William Laird, *Karluk; the Great Untold Story of Arctic Exploration*. 1976. Signet.

Mills, Stephen E., and Phillips, James W., *Sourdough Sky; Bush Flying Interior Alaska*. 1969. Superior Publishing, Seattle.

Niven, Jennifer, *The Ice Master; the Doomed Voyage of the Karluk*. 2000. Hyperion, New York.

Page, Dorothy G., *Polar Pilot; the Carl Ben Eielson Story*. 1992. Interstate Publishers, Danville, Illinois.

Page, Victor W., *Modern Aircraft*. 1930. The Norman W. Henley Publishing Company, New York. [How to unpack and assemble a JN-4]

Potter, Jean, *The Flying North*, 1945. The Macmillan Company.

Stevens, Robert W., *Alaskan Aviation History*, Vols I and II, 1990. Polynyas Press, Des Moines, Washington.

Swenson, Olaf, *Northwest of the World*. 1944. Dodd, Mead & Company, New York.

Taylor, John W.R., and Munson, Kenneth. *History of Aviation*, 1976. Crown Publishers, Inc., New York.

Tordoff, Dirk, *Mercy Pilot, the Joe Crosson Story.* 2002. Epicenter Press, Kenmore, Washington.

Wambheim, H. G., *Ben, In Memoriam. 1930.* 40-page booklet, printed in Hatton, North Dakota.

World Almanac and Book of Facts. 1995. World Almanac, Funk & Wagnalls Corporation.

Newspapers

Alaska Weekly, "Epic Career of Harold Gillam; Pioneer Alaska Air Ace Rivals Fiction." March 26, 1943.

Anchorage Daily News, Wright, Nancy, "Those Pioneer Aviators and the U.S. Mail," May 12, 1968.

Anchorage Daily News, Campbell, Larry, "Alaska Air Pioneer [William L. Lavery] Dies in Fairbanks," December 13, 1983.

Anchorage Daily News, Bell, Tom, "Crown Jewel of Alaska Aviation Comes Home," May 19, 1992. [the Wien Stearman].

Anchorage Daily News, McGee, Ron, "Museum Lands Part of History," August 16, 1995.

Anchorage Daily News, Bell, Tom, "Eagle River Man Who Won Medal in Rescue Dies," February 15, 1997 [Clyde Armitstead].

Anchorage Daily News, Obituary of Clyde G. Armitstead. Feb. 14, 1997.

Anchorage Daily News, Cox, Rose, "Aviator Joe Crosson Soars to Hall of Fame," Feb . 6, 2002.

Anchorage Daily News, Cox, Rose, "Eielson to Join Alaska Aviation Hall of Fame. March 15, 2003.

Anchorage Times, "Wien Air Founder Dies," July 19, 1977.

Anchorage Times, Gillette, Helen, "Widow Praises Pilot-Husband [Joe Crosson]," January 27, 1980.

Fairbanks Daily News-Miner, "Eielson Successfully Completes Trial Trip Air Mail to McGrath," February 23, 1924.

Fairbanks Daily News-Miner, "Wien Makes Trip to Eagle in Day Lands on River Bar," August 18, 1924.

Fairbanks Daily News-Miner, "Wilkins Assembling Vega Monoplane," March 1, 1928.

Fairbanks Daily News-Miner, "Wilkins Fails Discover New Lands in Flight to Spitsbergen," April 23, 1928.

Fairbanks Daily News-Miner, "Find Wreckage of Eielson Plane," January 27, 1930.

Fairbanks Daily News-Miner, Long, Everett, "What Became of Pilot Sigismund Levanevsky and N-209?" April 16, 1987.

Fairbanks Daily News-Miner, "Looking Back 50 years," September 16, 1987.

Fairbanks Daily News-Miner, Long, Everett, "Expedition to Canada Finds No Evidence on Levanevsky Crash," September 6, 1987.

Fairbanks Daily News-Miner, Bishop, Sam, "Two Expeditions Seek Fate of Soviet Flier [Levanevsky]," October 22, 1987.

Fairbanks Daily News-Miner, "Noel Wien: 1899-1977." July 19, 1977.

Fairbanks Daily News-Miner, Editorial, "Remembering Bill Lavery," December 12,1983.

Fairbanks Daily News-Miner, Long, Everett, "Pioneer Airplane Mechanic [Jim Hutchison] Recalls Pilot's Early Airborne Adventures," December 17, 1983.

Fairbanks Daily News-Miner, Long, Everett, "Russia, U.S., Join Forces for Rescue in Siberian Arctic," November 10, 1985.

Fairbanks Daily News-Miner, "Expedition Teams Prepare to Tackle Mystery of 'Lindbergh of North'" [Levanevsky], August 18, 1987.

Fairbanks Daily News-Miner, "Flying High Again," by Mary Beth Smetzer (restoration of Eielson's Jenny). May 24, 2009.

Appendix

TIMELINE, EIELSON/BORLAND SEARCH AND AFTERMATH

1929

Nov. 9 Eielson and Borland fly from Teller in Hamilton Metalplane. Did not return.

Nov. Frank Dorbandt repeatedly attempts to fly to North Cape with the Stinson Detroiter.

Mid Nov. Two dog teams, including that driven by Tzaret Berdieff, crewman of the *Nanuk*, make a ground search.

Dec. 18 Pilots Ed Young, Harvey Barnhill, Joe Crosson, Harold Gillam, all fail in an attempt to cross the Bering Sea due to weather.

Dec. 19 Joe Crosson and Harold Gillam cross Bering Sea to the *Nanuk* in a Waco 10 and a Stearman, both open cockpit biplanes.

Dec. 22 Harold Gillam with Stearman, Olaf Swenson as observer, flies a search near the *Nanuk*.

Dec. 23 Joe Crosson with Waco, and Olaf Swenson as observer fly a search east of North Cape for two hours plus.

Dec. 23 Harold Gillam with Stearman and Demetri Miroshnishenko as observer fly a search east of North Cape for two hours plus.

Dec. 31 Gillam's Stearman's landing gear damaged when engine quits on takeoff.

1930

Jan. Five dog teams from steamer *Stavropol* join search, authorized by Moscow.

Jan. 1 Joe Crosson makes twenty-minute test flight of his Waco.

Jan. 17 Joe Crosson flies Waco two hours plus, checking inland camp of Native reindeer herders.

Jan. 20 Gillam flies test flight of Stearman. It has been repaired by engineers from steamer *Stavropol.*

Jan. 26 Joe Crosson and Harold Gillam find lost Hamilton, crashed ten miles inland and ninety miles from the *Nanuk.*

Jan. 28 Canadian pilot Pat Reid, and Ed Young fly two Fairchild 71s from Teller to *Nanuk.*

Jan.29 Two Russian-flown Junkers arrive from Russian ice breaker *Litke.* Pilots are Commander Mavriki Slepnyov and Victor L. Galishev. Mechanics are Fabio Fahrig and Brednevya.

Feb. 3 A crew, of mostly sailors, from the steamer *Stravropol,* but including some local North Cape residents, starts chipping trenches in the four-feet-deep hard-packed snow, searching for the bodies.

Feb. 13 The body of Earl Borland is found in the hard-packed snow by digger volunteer T. Jakobson. It is within about four feet of the engine.

Feb. 18 The body of Ben Eielson is found lightly covered with snow 200 feet from the plane and 150 feet from where Borland's body was found. T. Jakobson is again the finder.

Feb. 23 Pat Reid flies from Teller with Fairchild CF-AJK loaded with gasoline, and breaks the landing gear while landing near the wreck of the Hamilton.

Feb. 23 Commander Mauriki Slepnyov flies the bodies of Eielson and Borland from the site of the Hamilton wreck, to the *Nanuk.*

Feb. 26 After examination by Dr. M.V. Kreszanev on the *Stavropol,* the bodies are taken by dog sled to Ed Young's Fairchild 71 NC153H, in preparation for flight to Alaska.

March 7 Delayed by bad weather, the funeral plane, Fairchild 71 NC153H, flown by Ed Young, and escorted by a Russian Junkers airplane and Harold Gillam's Stearman, lands at Fairbanks.

April 3 Joe Crosson flies mechanic Herbert Larison to the *Nanuk.* From there Larison travels by dog team to the broken Fairchild 71 CF-AKL at the Hamilton crash site. He digs under it to repair landing gear.

April 27 Joe Crosson flies Fairchild 71 NC9153 from Teller to the Hamilton crash site with mechanic Jim Hutchison and pilot S.E. "See" Robbins. Hutchison welds final repairs on gear of Fairchild 71 CF-AKL, and Crosson flies the plane to the *Nanuk.* Robbins follows with NC9153.

May 6 Both Fairchilds fly to Nome carrying between them forty bales of fur.

May 8 Joe Crosson, with S. E. Robbins aboard flies Fairchild NC9153 back to *Nanuk.* Crosson returns to Teller with twenty-six bales of fur. S.E. Robbins flies Waco that Joe used to find Hamilton from *Nanuk* to Teller with Larison as passenger. The fur-hauling contract is fulfilled, and all planes save the Hamilton are back in Alaska.

Index

Bold numbers indicate photos

Adler, Don (reporter), 49
Air mail flights to McGrath, 13, **14**, 15, 16
Aircraft
American Pilgrim 177, **202**. Described **208**. 209
ANT-6, N-209 Russian plane lost in Arctic, 176
Bellanca (diesel) crashes at Kotzebue, 129
Consolidated Fleetsters, 173, 174
Consolidated PBY (On search for Levanevsky) 177
DeHavilland DH-4, Military history of (footnote), 13
DeHavilland DH-4B Biplanes (Black Wolf Squadron), 6
DeHavilland DH-4BM Biplane, (Loaner to Eielson), 13, **14**
Fairchild 71 (Description) 140, Photos **152, 253,** 217
Fairchild 71 CF-AJK Damaged landing near wrecked Hamilton, 164
Fairchild 71 NC153H, Flies bodies to Fairbanks, photo **153H.** 165, 156, **182**
Fairchild 71s ordered for search. Assembled and test flown at Fairbanks 148. NC190H, One of three, crashes 148, 149. 188
Fokker *Alaskan* (Used for Wilkins expedition, 1926), *86*, 89, **90**, **91,** drops a wing, 96. In museum, 108, 109
Fokker *Detroiter*, Trimotor aka *Southern Cross*, **89**, 91, 92, **93**, 97,108, 109. 188
Fokker F.VII- 03m (Byrd's plane), 97
Fokker F-111, 33, 34, **35**, *36,* **38**
Ford Trimotor, **89, 233, 245**
Gage-Martin biplane, First airplane at Fairbanks, 6
*Golden Eagle,(P*rototype Lockheed Vega), 105
Hamilton Metalplane (NC10002), **56**, 55-*58, 59,* 78, **81**, **137.** Crash site found. Crash site described **151,** 152,153. Theories on crash 153,154. Cause of crash 159, 160. Pioneer Air Museum display of remnants **171**
Hisso Standard, On front cover. 23, **24,** 25, **26, 29,** 31, **32,** 38, **39,** 40
Josephine Ford, (airplane), 97
Junkers, 156, 157, 158, 160, **163**. Flies bodies to *Nanuk* 164, 165
Lockheed Electra, 197, 211, 212
Lockheed Vega NC-32M, *San Francisco*. Flown in Antarctica **118**
Lockheed Vega X3903 (description), **107**. In flight, **108.** 109. Flies to Spitzbergen, 109-112. 1,200-mile Antarctic flight 118. Mothballed in Antarctica, 119. Rots at Buenos Aires, 119
Los Angeles, (Lockheed Vega X3903 renamed) 117. Breaks through ice, 117. Antarctic flight, 118
OX-5 Jenny (NC47358), Front cover, **8**, 9, 32, 30, 31, 38, 171,172
San Francisco, Lockheed Vega (NC-32-M) in Antarctica, 116, 118

Southern Cross (airplane). In Canberra musum. See Fokker *Detroiter* Trimotor
Spad Fighters 216
Spirit of St. Louis, 93, 132
Stearman biplane (NC5414) (The "Wien Stearman"), 55, **58,** *59,* 60, 61, 142, 143. Used for search 144. 162
Stinson Detroiter #2 (C5262), 40, 41, 43, **45**, 47, 56, 60, *77,* 133, Flown to Nome, Fairbanks, burns, 143, 144
Stinson Detroiter (windblown on Lake Minchumina), **45,** 43-*47*
Stinson Detroiters, 40, 99, **100**,188, 189
Stinson Standard, **32**
Super Swallow biplane (NC2375), **37**, **141,** 100, 186, 189
Travel Air (Russell Merrill's), 53
Vega, Lockheed, 105, 106, **107**, **108**, 109
Waco 10 780E at Teller. Used for search 144
Waco biplane Model 9, (C2775), with OXX-6 engine, 52, 55
Winnie Mae, Lockheed Vega, 106
Zenith NC977Y (Gillam's), photo **205**
Alaska Aviation Heritage Museum, Anchorage, 58, 61, 142
Alaska Railroad, 5
Alaskan Airways, 127, 128, 133
Alexander, Charlie, 219
American Geographical Society, 88
Amundson, Roald, 97, 105, 113, 114
Anadyr (Siberia), 64
Anaktuvuk Pass, 94
Anchorage Air Transport, 55
Appleton, Tom (mechanic), 207, 208, 214
Armitstead, Clyde (mechanic, 173, 174, 175
Atlantic Aircraft Company, 33

Balchen, Bernt (pilot), 97, 131
Barbara Hernster (motorship), 63
Barnhill, Harvey (pilot), 143, 144
Barr, Frank (pilot), 214, 241
Barrack, James, 214
Bartlett, Captain Bob, 67
Barton, Jack, 219
Bassett, Bud, 133, 139
Baxter, Susan, 213
Bear, U.S.S. (Coast Guard) ship, 67
Beech, Walter, 131
Beederman, Horace (trader), 223
Belvedere, S.S. (former whaler), 66, 67, 71, 72, 77
Bendix Cup, 131
Bennett, A. A. (pilot) 31, 32, 36, 53, 183, 187, 206
Bennett, Floyd (pilot), 97
Berdieff, Tzaret (interpreter), 124, 140, On dog team search 150
Billberg, Rudy (pilot), 132, 207
Bissner, Bill, 124, 127
Black Wolf Squadron (at Fairbanks, 1920), 6
Blunt, Harry (pilot), 50
Borland, Earl Jr. (Bud), 167
Borland, Earl (mechanic) 7, 57, 60, 61, 134, 137, Body found 161. 163, 164 Funeral 167, 261
Borland, Irene (Earl's wife), 167
Borland, William, 166, 167
Branch, Harlee, 258
Brednevya (Russian mechanic), 158, 162
Brokhanov (Russian trapper), 161, 259
Brooks Range, 94
Brosius, Cal M. 93
Brower, Charlie, 177
Bunnell, Charles, UAF President, **160**
Burgenson, Charley, 228
Burke, Bill (mail team driver), 187
Burke, Paddy (pilot), 234, 235, 239
Byrd, Commander Richard E., 97

Carlson, Carl (miner), 186
Cathcart, Dan (pilot), 177, 203, 207
Cessna, Clyde, 131
Chelan, U.S.S. Coast Guard ship, 148
Chelyuskin (ship), 173, 174
Circle City, 219
Cope, Lon (pilot), 50
Crawford, Cecil (pilot), 38, 60
Crichton, Clark Jr. (cabin boy), 124, 134
Crichton, Clark (cook/steward), 124, 134
Cripe, Calvin "Doc," 78, **80**, *81*
Crocker Land, 88

Crosson, Joe (arrives Alaska) 36. With Super Swallow **37**. *38, 99*. In Antarctica 116, **117**. **141** *(photo)*. Arrives at Teller, 139. Awaits weather at Teller 140. 144. Flies to Siberia 145. Arrives at *Nanuk* 146. Flies on search 147. Test flight between storms and a two-hour search, 150. Final search Jan. 26. Locates crashed Hamilton 151. 156, **157**. In charge of digging for bodies 158. 159, **160**, Accepts bodies from Russians 165. 166, 169. Flies search for Levanevsky 177. The early years, 181. With Fairchild 71 (photo) **182**. In San Diego with JN4 183. Hired by Bennett 183. Flight to Teller 183, 184. First flights to villages 184. Flies Super Swallow biplane 184. Puts Waco 9 biplane into Moose River 184. Forced landing with Standard, 187. Afire aloft 187. 200 mile walk to Nenana 188. March 1927 flight Fairbanks to Barrow with Super Swallow 188-190. Return flight Barrow-Fairbanks with Swallow 190-191. Flies Fokker Universal float planes in east. In Antarctica with Sir Hubert Wilkins 192. Flies in Antarctic 192. Ferries Lockheed Vegas 182. Meets Wiley Post 192. Chief Pilot, Alaskan Airways 193. Arrives at Teller to search for lost Hamilton 193. Marries Lillian Osborne 193. Flies serum to Barrow in Wien Stearman and Fairchild 71, 194. Repairs Wiley Post's Winnie Mae, 194. Mercy flights 195, 196. Flies bodies of Rogers and Post to Seattle, *196*. Photo with Post and Rogers, **197**. Refuses medals 198. Division Manager for PAA. Death in 1949, 258, 259

Crosson, Marvel, 181. Learns to fly 183. Arrives Fairbanks 191. Wins Air Race in California 192. Killed in woman's air derby 193

Curlee, Sgt., 221

Curtiss Northwest Airplane Company (flight school), 20

Cutting, Percy, 213

Danforth, Marcel L. "Danny," 204

Dayo, Mrs., 34

De la Vargne (Fairbanks mayor), **160**

Deception Island, 117, 118,119

DeHayes, Floyd Andrew (legal name of Frank Dorbandt), 235

Delon, Philip Father, 129

Detroit Arctic Expedition, 38, 88

Detroit Aviation Society, 88

Dodson, James (pilot) 214

Dog team mail, and contracts 10, *11*

Doolittle, Jimmie, 131

Doonerak (Mt.), 94

Dorbandt, Frank (pilot), 55, 133, 134, 135, 135. Challenges Eielson to fly 136. Damages axle of Detroiter, 138, 139. Awaits weather at Teller,140. 228. Remembered by Frank Barr 232. Flies Ford trimotor Seattle-Alaska **233**-240. Arrives Anchorage with Ford 240. Grounded by C.A.A. at Anchorage 240. Rescues injured fisherman 240. Flies Father Hubbard 242. Prop breaks wrist 242. Dies of pneumonia 243

Downer, Midge, 34

Draven, Arnold (Second Mate), 124

Dunlap, Carl C. (gold miner), **29**

Earhart, Amelia, 106

Edwardson, Tony (trader), 103

Eielson, Ben, as second Lieutenant **3.** Education, military career, teacher at Fairbanks, with Capitol Police 3, 4, 6. In OX-5 Jenny with passenger **8.** Airmail flight to McGrath, receives gold watch 15. With Fokker *Alaskan*, **85**. Outlines Wilkin's exp goals, 90. Crashes Fokker *Alaskan*, **91**, 92. First flight Fairbanks to Barrow, 94. Pilot in Florida 99. 1927 flight over Arctic Ocean. First landing ever on ice of Arctic Ocean100. Crashes Stinson *Detroiter #1* on Arctic Ocean 101. Walks withWilkins on ice to Beechey Point 102, 103. Works for Bureau of Aeronautics, **104**. Exploits

with Lockheed Vega X3903, *107.* Honored in Europe and U.S., 112-115. In *Antarctica* **117**, 127. General Manager Alaskan Airways 128, 129. Checked out in Hamilton 132. Damages Hamilton landing gear 133. Lands at *Nanuk* 134. Arrives Nome with fur in Hamilton 135. At Teller awaiting weather for flight to *Nanuk*. Departs Teller 136, 137. 164. Body found 162, funeral **167**, 168, memorials **170**,171, 172
Eielson, Ole, 166
Elisif (Schooner), 69, **70**, 71, *72*, 75, *77*, 78, *80,* 81, 123, 124 (sinks), 125, 126, 241
Ellsworth, Lincoln, 97
Exposition Park (footnote), 8

Fahrig, Fabio (Russian mechanic), 158, **160**, 163, 164
Fairbanks Airplane Company, 30, 50, 183
Fairbanks Daily News-Miner, Aviation Notes, 48, 137, 138
Fairbanks, description, 5
Farnsworth, Ira, 7, 8
Farthest-North Airplane Company, 7, 12
First Alaska airmail service contract, 12
First Alaska scheduled flights, 42
First Fairbanks to Nome flight, 34, 36
First flight from North America to Asia, 82
First flight over unknown of Arctic Ocean, 96
First Interior Alaska commercial flights (by Eielson) with Jenny, 9
First mail flight (by Eielson, in 1923), 7
Flight to Siberia (Wien), 77
Fordson tractors (Snow Motors), 93
Fort Yukon (described), 221
Fox Film Expedition, 52, 53
Fox furs, Siberia (value) 64
Franzen, Ernie (mechanic), 38, 187

Galkovsky, Nicoli (Russian radioman), 176
Galyshev, V.L. (Russian pilot), 150, 158
Gaston, William (radioman), 116
Gebo, Robert, 213
Geist, Otto, UAF palaeontologist, **160**
Gillam Airways, 207
Gillam, Donald, 214
Gillam, Harold Jr. 214
Gillam, Harold, *142*, 143, 144. Flies to Siberia 145. Arrives at Nanuk 146. Flies Swenson on search 146. Flies two hour search 147. Damages Stearman in forced landing 147. Flies final search Jan. 26, 150. Locates crashed Hamilton 151. 156, **157**, 158, 159, **160**, 162, 163, 164, Accepts bodies from Russians, 165. 166. Flies support for Levanevsky search 177. Flies to Siberia 193. Photo with Wien Stearman **201**. Description and early years 203. Buys Jenny 204. Has flying lessons 204. Survives crash 204. On Eielson/Borland search 204. At Copper Center 204. Flies trader McCrary to doctor 204, 205.Flies miner Carl Whitham to doctor, 206. Passes airman's exam 207. At Fairbanks with American Pilgrim planes 207. Photo in Pilgrim, **208.** Weather Bureau contract 209. Kuskokwim mail route 210. Chief Pilot for Morrison-Knudson 211. Crashes Lockheed Electra 212. Search for Gillam 212, 213. Body found 214. CAA report on crash 214. Fairbanks Funeral 214. 229
Gillam, Maurine, 214
Gillam, Wenona, 214
Glass, Don (pilot), **233**, 234, 235, 236, 237, 238
Gleason, Robert, 123, 124, 133, 134, 119, 144, 149, 158, 159, 169, 197. Early years, 256, 257. Engineer for Pan Am in Alaska, 257. Returns to Siberia with Crosson, 257, 258. With Air Force in WWII, 258. Postwar with Pan Am in Alaska, 258. With Aeronautical Radio Inc. 258. Book *Icebound in the Siberian Arctic*, 258
Goldstein, Charlie (fur buyer), 50

Gotherberg, Fred "Mulligan" (trapper), 50
Graham Land, 118
Graham, Alger (pilot), 99. Flies Stinson Detroiter over Arctic Ocean 1927. 104, 188
Gunnison, Silas, 219

Hale, Martin, 235
Hall, Murray (CAA), 240
Harding, President Warren, at Fairbanks, 9
Hart, Billie (secretary), *27*
Hatton Aero Club and Jenny, 4
Hawks, Frank (pilot), 106, 131
Hearst, William Randolph, 116
Hectoria (Norwegian whaling ship), 116, 117
Herbert, Lyle, 50
Herman (whaling ship), 67
Hines, Art, 228, 229
Hisso engine unsuited for deep cold, 33
Hobby (Norwegian ship), 113
Holman, Charles W. ("Speed"), 4, 131, 246
Holmstrom, (Mate), 124, 134
Hoover, President Herbert, 166
Hubbard, Father Bernard J., 242
Hubbard, Percy, 228
Hudson, Eddie, *24*
Hufford, Andy, (mechanic), 38
Hughes, Bill (Canadian mechanic), **160**, 166
Hummel, William "Hosie, the High-powered Swede," 16
Hunter, George, 124
Huntley, Charles (radioman), 75, 123, 125
Hutchinson, Palmer (reporter), 88, 90, 91
Hutchison, Jim "Hutch" (mechanic), 97, 149, 150, 174, 189, 218, 259. Early years, 260. Works for explorer Wilkins, 260. Works on *Winnie Mae*, 260. Photos of **257** *260.* Makes heater for Super Swallow, 261. Awards, 261. Hutchison Career Center, 261. Death of, 261

Ingraham (mining engineer), *27*

Jakobson, T., Russian digger, 161, 162
James Gordon Bennett Cup, 131
Jochimsen (Arctic boat pilot), 125
Jochimsen, A. P., 241
Jones, Roy (pilot), 34

Kamasgaard's roadhouse, 43
Karise (ship), 241
Karluk (ship), 67, 87
Kavelin (Soviet rep at North Cape), 71, 72
Keenan Land, 88
Kendrick, Sam S. (Reindeer Service), 50
Kester, Marshal, 234
King & Winge (ship), 67, 68
King, George (mechanic), 186
King's Bay, Spitzbergen, 97
Kirkpatrick, M.D., 207
Knight, "Tiny," 223
Knox, Bill (pilot), 197
Krasin (Russian icebreaker), 176
Kreszanev, Dr. M.V., examines Eielson and Borland's bodies, 165

Lanphier, Captain Thomas G. Jr., 88
Lanphier, Major Thomas G., 88, 91, 92
Larison, Herb (mechanic), 168, 169. Repairs Fairchild 71 CF-AJK in Siberia, 259. 261
Larsen, Evan (captain of *Elisif*), 69, 70, 125, 126
LaVern, Doctor, 228
Lavery, Bill, 174, 175
Leach, Frank, *26*
Leach, Pop, 220
Levanevsky, Sigismund (Russian pilot), 173, 175-178
Lien, Bert, 207, 209
Lincoln, Nicholas, Sr. (recollects Gillam), 206, 207
Lindbergh, Charles, 93, 129, 130, 131, 132, 197
Litke (Russian ice breaker), 147, 156
Lomen, Alfred J., 143, 147
Loughheed, Allen (airplane builder), 106

Martin, James, 6
Martin, Lily, 6
Mattern, Jimmy, (pilot),176
McCafferty, Frank (mechanic), 13
McCauley, Sam (Canadian mechanic), **160**, 166

McCrary, John (trader), 204, 205, 206
McCrary, Nels, 20
Meals, Owen, 78
Meherin, Joseph J., *26*, 50
Memphis, U.S.S., 131
Merrill, Russell (pilot), 53
Metzdorf, Dewey, 213
Miller, Major Ray S. (pilot), 20
Moore, Ed (mechanic), 143, **257**
Moore, Johnny, 207
Moore, Leo, 207
Morrill, E. W. (pilot), 22, 23
Morrison-Knudson Company, 211

Nanuk (schooner), 69, 71, 72, 123, 124, 126, **127**, 133, 134. history after 1932
National Air Races, 131
Nelson, "Phonograph," 222
Newhall, Dr. (amputates Eielson's finger), 103
Nickoloff, Peter, **160**
Nieminen, Matt (pilot), *32*, 50, 53, 149, **160**
Nimrod (dentist at Eagle), 223, 224
Nobile, Umberto, 97
Nordale, Hjalmar, 218, 229
Nordale, Mrs. Ladessa, *8*,
Norge (dirigible) 97, 183
North American Newspaper Alliance, 88
North Cape, Siberia (village and geographical location), 70, 78, 133
Northeastern Siberian Company Limited, 62, 63, 65
Northland, (Coast Guard ship), 125
Northrup, Jack (airplane designer), 106
Northrup, Marvin (airplane dealer), 23
Nystrom Charles, 16

O'Conner, Jack (wildlife agent), 241
Okpeaha, Claire, 197
Ottilie Fjord (early name of *Nanuk*), 69, 74

Pacific Alaska Airways, 50
Palm, Johnny, 218
Pan Am chooses Pacific route, 258
Parker, Genevieve, 90
Parmenter, Elbert, 206
Peary, Robert, 105
Peck, Lyman, 258
Pedersen, Captain C. T. (whaler captain), 67, 71
Peterson, Ray (pilot), 210
Pioneer Air Museum, Fairbanks, 95
Podgorsky, Count, 63
Polister, Ray S. (VP Swenson Fur Co.), 75, **80**, *82*, 125
Pollack, Frank (pilot), 228, 240
Pope, R. A. (auto manufacturer), 38
Porter, Orval (mechanic), 116, **117**, 168, 191, 218, **257**
Post, Wiley, 106, 131, 195, 196. **197**, 198, 237
Powers, John (senator), 223
Powers, Ma, 223
Pratt & Whitney Wasp (engine), 57
Prudhoe Bay oil fields, 94

Quigley, Fannie, 185, *223*, 228
Quigley, Joe, 184, 185

Radio Corporation of America, 75
Reeve, Bob (pilot), 54
Reid, T. M. (Pat) (Canadian pilot), 140, 148. Forced landing of Fairchild 149. Arrives at *Nanuk*. **157**, 158, 159. Breaks Fairchild 71 while landing in Siberia 164, 165, 166
reward offered by Russians for word of Hamilton, 148
Rickarts, Paul (footnote) 8
Robbins, S. E. (pilot), 50, 169, 228, 259
Rodebaugh Fairbanks Airplane Company, 38
Rodebaugh, Jimmy, 17, buys airplanes, hires pilots and mechanic, 23, *24*, *32*, 33, 36, 183
Rogers, Will, 196, **197**, 198
Romaker's roadhouse, 219
Root, Billy, 218
Ross, Hosea (undertaker), 229
Ruby, Yukon River village, 36

Salmi, Charlie (miner), 186
Samson, Art (pilot), 24
Schiek, Charlie (carpenter), 217
Schmidt, Dr. Otto, 173
Schneider Trophy, 131
Seppala, Leonard, **32**, **197**
Seppala, Sigrid, **32**,
Shafer, George (Gov. of N. Dakota), 168

Siberia (map), 85
Siberian Eskimos, 65, 66
Slepnyev, Mavriki (Russian pilot), 150, 158, 159, **160**, 163, 164, 164, 165, 166, 173
Smallpox epidemic, northern Alaska, 55
Smith, A. M. (reporter), 99, 116, 188
Smith, Charles Kingsford (pilot), 109
Smith, Clyde, 214
Snow Motors, 93. 94, **95**
Southern Cross (ocean liner), 116
Soviet Fur Trust, *69*
Spitzbergen, 97
Springbett, Gordon, 218, **257**
Star Route Air Mail Contracts, 43
Stavropol (ship), 124, 127, 147, 148, Crew repairs Wien Stearman, 150. 157, 158, 161, 165, 165
Stearman, Lloyd, 131
Steel, Mr. & Mrs., 223
Stefansson, Vilhjalmur, 18, 66, 87
Stines, Norman C., 34
Sutherland, Dan, Alaska's delegate to Congress, 5, 12
Sutherland, Dr. J. A., 55
Swartzman, G. (Canadian pilot), 143, 148
Swenson Fur Trading Company, 68
Swenson, Marion, **74**, 123, 133, 158, 159, 167
Swenson, Olaf, 62, 63, 67, 68, **74**,123, 124, 127, 133, 138, 140, 158, 159, 167
Swenson-Herskovitz Company, 77

Talcott, Frank, 4
Thach, Robert, 258
Thiele, Carl (acting Alaska governor) 147
Thompson Trophy, 131
Thompson, Charles (trader), *223*, 224
Thompson. W. F. ("Wrong-font"), 7, 8, 48, **49**
Tilman, Warren O. "Tillie," 229, **230**, *231*, **257**
Tippets, Joseph 213
Tobin, Jack, 17
Tondro, Frank L. ("Malemute Kid"), 37
Travel Air Company, 131
Treacy, George (passenger), 40
Turner, Roscoe, 131

U.S.S. Bear (Coast Guard ship), 67
Uelen (Siberian village), 125
Ulm, Charles (pilot), 109

Walsh, Father W.F., 129
Warren, Helen, 135
Warren, Jack, 135, 137, 218
War-time emergency landing strips, 211
Weeding, R. H. (captain), 124
Weeks Field (footnote), 8
Weeks Field, 218
White, Sam O., 129, 228, 229, **253**, 255
Whitham, Carl (medivac by Gillam) 206
Wickersham Dome, 31
Wien Alaska Airways, 52, 75
Wien, Ada, *59*. With Noel (photo) **130**
Wien, Fritz, *61*
Wien, Jim, **59**
Wien, Julia (Ralph's wife), 59
Wien, Noel Merrill (Noel and Ada's son), 29
Wien, Noel, meets Eielson 17. Early life, learning to fly, barnstorming 19, 20. Arrives Anchorage, 25. Flies to Fairbanks 25–29. With Hisso Standards at Fairbanks in 1924, **24**. Flies for Rodebaugh at Fairbanks, 30-36, 49-50. Forced landing with Hisso Standard, 31. With Fokkeer F III, 35. Based at Nome, 38-42. Establishes Nome-Fairbanks flight schedule, 42–47. Alaska's first bush pilot, 50. With family, **59.** Decides on Siberian flight, 75. In Siberia 70-82. With Hamilton, **83,** Witness of two Fokker crashes, 92. Flight to *Elisif* 128. Leaves Alaska temporarily 244. With wife Ada (photo) **130.** Air time in Hamilton 132. Checks Eielson out in Hamilton 132. Travels Outside in 1929, 244. Buys Stinson Jr. NC490H, 244. Returns to Alaska to fly for Alaskan Airways, 245. Founds Wien Airways of Alaska, 246. Spends winter night in cold cabin, 247.

First commercial passenger flight, Fairbanks to Seattle, 248, 249, 250. buys Ford Trimotor, 250. Goodale portrait of Noel, **251**. Flights with the Ford Trimotor, 252, 253, 254. Photo with Sam O. White, **253.** Building an airline, 254. Honors, 254, 255. Death, 255.
Wien, Ralph, 34, 36, 38, 54, 55, 57, **59,** 61, 77. Killed at Kotzebue 129, 217.
Wien, Richard A. (Noel's son), 29, 235
Wien, Robert (Ralph's son) **59**
Wien brothers build hangar at Weeks Field in 1929, 54
Wilkins, George Hubert, 18, 40, 41, 87, **93.** Measures depth of Arctic Ocean 100. walks with Eielson to Beechey Point 102, 103. Photo of **104**. Flies over Arctic Ocean in Stinson Detroiter #2. With Lockheed Vega X3903, **107**. Honored in Europe and U.S., 112-115. Knighted as Sir Hubert Wilkins, 114. In Antarctica, 116–119. Career (condensed), 119, 120, at Eielson funeral 168. Searches for Levanevsky 176, 177
Williams, Frank, 236, 239
Wimmler, Norman (mining engineer), 49
Winchell, Oscar, 206, 207
Winter Flying methods by Wiens, 51
Wood, Dick, 7, 8, 9, 12
Woods, Earl, 207
Wrangel Island, 67
Wright Whirlwind engine (footnote), 41
Wright, Orville, 38
Yenukidze, A.S. (USSR administrator), 175
Young, Ed (pilot), 32, 50, 143, 144, 149,156, **157**, 158, 159, 163, accepts bodies from Russians, 165. Flies bodies to Fairbanks 166. 185. 187. as Army pilot in WWI 216. As hunter/trapper in Alaska **216**. Mail flights, Alaska 217-228. Faces demanding trapper 220. Snatches girl from prop 221. Skill as pilot 225. emergency gear 226. With Fairchild 71 (photo) **227**. Maintenance Engineer for P.A.A. 228. Killed at Livengood 229.
Yunker, Bill (mechanic) 24, 28

About the Author

Jim Rearden arrived in Alaska in 1947 to work as a summer Fishery Patrol Agent at Chignik for the U. S. Fish and Wildlife Service. In 1950 he organized the Wildlife Department at the University of Alaska, Fairbanks, where he taught as head of that department for four years. He resigned to become a free-lance outdoor writer and photographer. To accompany that profession he became a registered big game guide.

To support his writing, he also worked as a carpenter, an office manager for a construction company, a clerk in a trading post, and as a commercial salmon fisherman. From 1959 through 1969 he was Area Biologist for Commercial Fisheries for Cook Inlet for the Alaska Department of Fish and Game. In 1970 he became the Outdoors Editor for *Alaska Magazine*, as well as a Field Editor for *Outdoor Life* magazine, and held both positions simultaneously for twenty years.

He has written more than 500 features for more than forty magazines in seven countries around the world; *Alaska's First Bush Pilots, 1923–30* is his 27th book on Alaskan subjects.

He served on the Alaska Board of Fish and Game and the Alaska Board of Game 1970–82. In 1976, President Gerald Ford appointed him to the National Advisory Committee on Oceans and Atmosphere where he served for eighteen months.

He is a veteran of WWII, and was a sonar operator aboard the *U.S.S. Lovering,* a U.S. Navy destroyer escort in the Central Pacific war zone. He holds a private pilot's license, and has owned three airplanes. He holds degrees in wildlife conservation from Oregon State University and the University of Maine. In 2005, in recognition of his teaching, wildlife conservation career, and writing, he received an honorary doctor of science degree from the University of Alaska, Fairbanks.

With his wife Audrey, he lives in Homer in a log house he built himself.